Island of

Island of Woods

How Ireland Lost Its Forests and How to Get Them Back

Paul McMahon

NEW ISLAND

ISLAND OF WOODS
First published in 2023 by
New Island Books
Glenshesk House
10 Richview Office Park
Clonskeagh
Dublin D14 V8C4
Republic of Ireland

www.newisland.ie

Print ISBN: 978-1-84840-879-1
eBook ISBN: 978-1-84840-880-7

Typeset by JVR Creative India
Edited by Neil Burkey
Cover design by Anna Morrison / annamorrison.com
Printed by FINIDR, Czech Republic, finidr.com

New Island Books is a member of Publishing Ireland.

10 9 8 7 6 5 4 3 2 1

To my son Dylan and all our walks in woods together

Forest cover in Ireland

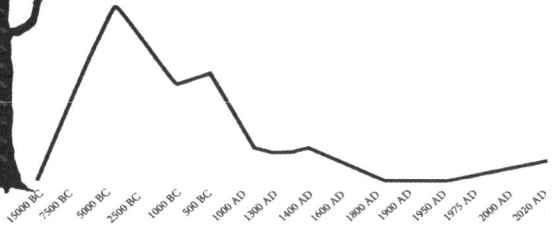

100%
90%
80%
70%
60%
50%
40%
30%
20%
10%
0

15000 BC 7500 BC 5000 BC 2500 BC 1000 BC 500 BC 1000 AD 1300 AD 1400 AD 1600 AD 1800 AD 1900 AD 1950 AD 1975 AD 2000 AD 2020 AD

CONTENTS

Introduction   ix

1. Going Native   1

2. Early Humans   15

3. Celtic Ireland   31

4. Medieval Invaders   51

5. Conquest and Commerce   68

6. The Two Irelands   88

7. Reforesting Ireland   109

8. Grinding to a Halt   130

9. A Sylvan Future?   149

    Acknowledgements   170

    Sources   172

    Index   198

INTRODUCTION

WHEN MY FATHER DIED, I spent a few days sorting through boxes in our old family home in Dublin. In one of the boxes I found a curled photograph of the view from my bedroom window. In the 1980s we lived in a dull suburb of identikit houses, each with its garage and garden, but we were only a short bicycle ride to where the city banged up against a more ancient countryside. From my window, past the rows of houses, I could see the green folds and fields that led up to the Dublin and Wicklow mountains behind. At that time the mountains were mostly bare, their contours visible. There was a faint hint of recently planted trees on some of the hillsides, but they were too small to stand out. The only mature woodland was a small band that formed part of the old demesne of Cabinteely Park, enfolding a great house built in the 1700s.

The view looks very different today. The trees on the mountains have grown and the backdrop to the city is now dominated by plantations of Sitka spruce, a fast-growing conifer that comes from the Pacific coast of North America. The plantations form large, angular blocks of dark green that never lose their colour, even in winter. They have grown so tall that they have altered the horizon. Where the plantations start and end, there is an abrupt step in the skyline, as if a giant had laid down blocks on top of the mountains. Where groups of trees have been felled within the blocks, there is a gap in the silhouette, like a missing tooth in a boxer's grimace.

The conifers planted in the 1970s are now ready for harvest. Under the standard Irish forestry model, when commercial forests reach maturity they are cleared, the timber sold and a new crop of trees planted for the next rotation. But when Coillte, the state-owned forestry company, announced plans to do this in 2017, all hell broke loose. The problem was that the mountains had become an important recreational resource for Dubliners, with over 600,000 visitors per year. These visitors didn't like the idea of giant machines laying waste to the landscape, even if, paradoxically, many had grumbled about the appearance of these non-native conifers in the first place. People, rightly, feel a sense of ownership over their landscape, and they are sensitive to change, especially when it involves cutting down trees. There were angry letters to *The Irish Times*, and the Green Party put out a campaigning video. It was the most visible of issues, as a million Dubliners could look up and see what was happening on the hillsides above them. Coillte was forced to back down, change its plans and embrace a different kind of forestry.

The furore over clear-felling in the Dublin Mountains is just one of the controversies to affect Irish forestry over the last few years. Woodland management excites such strong opinions in Ireland that it can bring crowds out onto the streets (albeit small ones). In 2019 campaigners from Leitrim staged a protest outside Ireland's parliament, the Dáil. Chanting 'soils not spruce' and 'communities not conifers', they called for an end to the widespread planting of Sitka spruce in the county, claiming that it damaged the environment and displaced local farmers. They advocated a more sustainable forestry based on native species and community ownership. Two years later angry foresters, carrying chainsaws instead of pitchforks, staged their own protest outside the Dáil. They lambasted the government for withholding the licences they needed to plant and harvest trees, warned that the forestry sector was in crisis and pointed out the importance of commercial forestry for storing carbon and supplying materials. Forestry, it seems, is front-page news.

The expansion of Irish woodland is grinding to a halt, mired in conflict between foresters, businesses, farmers and environmentalists. And no one can agree on how to manage the forests that already exist. This reflects a deeper ambivalence towards forests within Irish culture. Although we may cherish individual trees, many of us are attached to a 'traditional' landscape of open fields, trimmed grass and hedgerows. Farmers are reluctant to give up a way of life centred on rearing cattle or sheep, even if there is little profit in it. And environmentalists, who you would normally expect to be the biggest tree-lovers, have emerged as the most vocal opponents of the current forestry model based on imported conifers rather than native broadleaves.

This book will try to explain how we ended up in this situation. It takes a long historical view, starting when the ice sheets retreated more than 10,000 years ago, leaving behind a treeless landscape. Trees quickly recolonised Ireland from other parts of Europe, creating an island of woods. Eight thousand years ago the land behind my old bedroom window would have been cloaked with oak, ash, hazel and birch, rising close to the mountaintops. But not all trees made it back across the sea after the last glaciation. This has led to distinctions between 'native' and 'non-native' species, and to a sort of arboreal apartheid, where some species are valued higher than others. But what exactly does 'native' mean? Are native species *better* than later imports? How do we reconcile our fondness for native broadleaves with the economic realities of commercial forestry built around imported conifers?

By the beginning of the twentieth century, following centuries of woodland clearance, the island was practically treeless again. Only 1 per cent of Ireland was covered in woodland, by far the lowest amount of any European country. How did this happen? We Irish like to think of ourselves as tree-lovers. We have a romantic notion of a Celtic sylvan past, something that lives on in a tradition of sacred trees dotted around the countryside. The corollary is that we tend to blame foreigners for destroying our arboreal heritage. The popular view is that it is all the fault of Queen Elizabeth or Oliver Cromwell, collateral damage in the bloody English conquest of Ireland 500 years ago. Is this true? The following chapters tease out the reasons behind the disappearance of Irish forests. They show how the interplay of ecological, social and economic factors shaped the Irish landscape over thousands of years.

Over the last century this process of deforestation has been reversed. Because of one of the most ambitious tree-planting programmes in Europe, Ireland's forest cover has grown tenfold. Forests now cover almost 11 per cent of the island. This represents a major transformation of the Irish landscape. How was this achieved? The book explores the people and the ideas behind the great reforestation of Ireland that began in the early 1900s and reached its peak in the 1990s. It points out the successes and the failures, and the reasons why Irish forests look so different from those in the rest of Europe. The nature of this reforestation effort also helps explain why attitudes are so polarised now.

Forestry in Ireland today is logjammed, beset by conflicts and disagreements about which trees to plant, where to plant them and how to manage them. This is a shame, as there are good reasons to want more trees in Ireland. Forests suck carbon dioxide out of the atmosphere, lower temperatures and help avert the looming crisis of climate change. Forests produce the wood products we need for building, energy and materials, creating jobs and wealth in a sustainable circular economy. Forests also provide a place for recreation and mental restoration: everyone knows that a walk in the woods is good for the soul. Drawing on the experience of other European countries, and inspired by Irish innovators who are already applying new ideas, the final chapter of this book points to a more sustainable form of forestry that can reconcile the competing interests and deliver the multiple functions – economic, social and environmental – that we need from our woodlands. We are about to enter a new age of wood, so we need to get this right.

This book is about the past, present and future of Irish forestry. But there is a wider relevance to this story. In an attempt to slow climate change, many countries have adopted ambitious tree-planting programmes. Even without this, landscapes are changing. As economies develop, people are drifting to the cities, giving up traditional agrarian activities and abandoning less productive agricultural land, allowing room for forests to expand, whether by design or neglect. Ireland's modern reforestation shows what can be achieved with a concerted policy in a relatively short period of time; yet it is also a cautionary tale of how landscape change can spark resistance among different parts of society, while also causing unintended ecological consequences that are hard to fix.

My own journey has taken me from Dublin to London and New York and a few places in between. After studying history, I helped set up a business that invests in ecological forms of land management in Ireland and elsewhere. This gave me a front-row seat to observe the conflicts and contradictions of Irish forestry. I have had the privilege of working with pioneering Irish foresters who have spent their careers trying to forge a more sustainable path. My experience prompted me to dig deeper into the dynamics that shaped Irish woodlands in the past. This work is an attempt to explain why the Irish landscape, and the view from my old bedroom window, looks the way it does now. And it offers some suggestions for how we can shape the Irish landscape over the next hundred years, so that future generations sorting through boxes of old photographs can look up and thank us for the legacy we left.

GOING NATIVE

TWENTY THOUSAND YEARS AGO Ireland was a country with no trees. Looking down from space, you would not have seen a fleck of green. Instead, vast sheets of ice, one kilometre thick, covered the entire island and extended out onto the continental shelf. These blue-white sheets stripped away the soil and ground up the bedrock, carving out U-shaped valleys, such as Ireland's only fjord, Killary Harbour in Connemara. Ice, and the meltwater flowing beneath it, deposited gravel and sand in long sinuous ridges called eskers, or scooped it up into elongated hills called drumlins, which lie mottled across the Irish countryside like half-buried eggs. Ireland has such an ice-carved landscape that a number of glacial terms used around the world were first coined by geologists there: esker derives from the old Irish *eiscir*, which means 'ridge or elevation', and drumlin comes from the Irish word *droimnín*, meaning 'little ridge'. There was no place for trees (or humans) in this icy terrain.

After the Last Glacial Maximum the world began to warm up. The ice sheets slowly receded to the north and east. They left behind a devastated landscape of exposed rock and bare soil. But nature has extraordinary healing properties. It fills any niche. First came algae, mosses and lichens, along with the bacteria, fungi and insects that do the hard work of building soil. Then came sedges and dwarf shrubs such as juniper (*Juniperus communis*), which are characteristic of arctic tundra and high alpine slopes. If you went for a walk in the midlands at this time you might have seen Irish giant deer grazing on these plants – their massive antlers, four metres across, can be seen in museums across Ireland.

By 15,000 years ago Ireland was ice-free. There were no large trees yet, only woody shrubs, but natural succession was heading in that direction. Progress was interrupted by a cold snap 12,300 years ago, which brought more snow and ice, reducing the vegetation to low tundra plants. Species such as the Irish giant deer could not adapt quickly enough and became extinct. But by 11,700 years ago the cold had finally gone. The climate warmed rapidly, shifting from arctic to temperate conditions within a decade, and stabilising at a point where Ireland was 1–2°C hotter than today. The stage was set for the arrival of the tree species that are now considered native to Ireland.

Irish scientists have spent decades digging around in bogs and studying ancient pollen and half-preserved timber in an effort to piece together the story of how forests became established. We have a good idea of which trees colonised Ireland and when. However, the mechanism by which they arrived on Irish shores is still a puzzle. And more recent research raises questions about which trees are 'native' and which are 'foreign' – and what this really means.

THE FIRST TREE TO colonise Ireland was birch (*Betula pendula* and *Betula pubescens*), a thin-leaved, deciduous hardwood with triangular, serrated leaves. It seems to have appeared all across the island, all at once. Birch is the ultimate pioneer species. Wherever there is bare soil or abandoned open ground, you can usually count on birch being first to the scene. It hates shade and races to get to open spaces. Because it produces lots of tiny winged seeds that disperse on the wind, it usually wins the race. Birch grows fast, up to 1 metre per year, and can set seed within ten years. It does not grow very high, and is usually replaced by taller, more patient trees through a natural process of succession. Birch is a pesky tree for foresters in Ireland today. Although it makes good firewood, it does not have a high commercial value. It tends to invade failed forest plantations, abandoned farmland or land that has been cleared of trees. However, it can also act as a 'nurse' tree for other species, helping them get established before making way.

Birch played nurse to the next tree to arrive in Ireland – hazel (*Corylus avellana*). Hazel is a medium-sized tree that can grow to 12 metres in height. It has a smooth, grey-brown bark that peels with age, and its leaves are soft to the touch as a result of the downy hairs on the underside. Hazel is well known for its yellow 'lamb's tail' catkins in spring, and its edible nut has been enjoyed by people and animals for millennia – the Nutella sandwich is just the latest incarnation. Hazel first appeared in Ireland around 9,500 years ago. As the climate warmed and dried, hazel grew up between birch and eventually overwhelmed it, forming dense forests in the west and north. For a while, it was the dominant tree in Ireland: its pollen from this period is seventeen times more abundant than all other tree species put together.

Today hazel can be found scattered through hedgerows and woodlands across Ireland – its nuts are a forager's treat in autumn. And there are patches of hazel woodland in the Burren in County Clare that provide a glimpse of what the landscape looked like in the postglacial era. But hazel trees are not grown commercially on any scale.

Hazel's days of glory were numbered, because it soon faced competition from three forest canopy species that would define the Irish skyline for thousands of years. The first was pine, which appeared on Irish shores around 10,500 years ago. There are more than 100 species of pine, but the one that made its way to Ireland was Scots pine (*Pinus sylvestris*). Despite its Scottish name, *Pinus sylvestris* is the most widely distributed pine species in the world and can be found all across Eurasia. It is an evergreen conifer that grows up to 35 metres in height. It has short, blue-green leaves and an orange-red flaky bark. The mature tree has a distinctive appearance, especially when silhouetted on top of a hill. Its lower branches fall off, leaving a long, bare and straight trunk, capped by a flat-topped mass of foliage. Scots pine grows moderately fast and produces high-quality timber. It is grown commercially today, although it has lost place to other faster-growing conifers. It has had a complicated and elusive life in Ireland over the last 9,500 years, as we will see later.

Around 9,000 years ago Scots pine was joined by another large tree, this time a hardwood species – the elm (*Ulmus glabra*). The central part of elm in the history of Irish forestry may come as a surprise. It is hard to find a mature elm tree in Ireland today. In the 1970s most were wiped out by Dutch elm disease, a fungus originating in Asia and so named because it was identified by Dutch scientists. When healthy, the elm can develop into a magnificent forest tree, up to 40 metres in height. It has oval-shaped leaves with jagged tooth edges, which grow along the stem in a zigzag pattern. Elms are rarely planted now. They are sometimes found growing in graveyards in Ireland and are associated with death, perhaps because the trees can drop dead branches without warning. In the past, most coffins were made from elm wood.

If the elm has receded in the Irish imagination, the same cannot be said for the iconic oak. The oak tree was traditionally a symbol of strength, kingship, endurance and fertility, and is often considered the king of trees. It conjures up misty visions of a sylvan Celtic past. Oaks appeared in the south of Ireland at the same time as the elms, around 9,000 years ago. Oak trees are one of the largest and longest-living broadleaf trees in Ireland, growing to heights of up to 40 metres and capable of living more than 1,000 years. There are two species native to Ireland: the sessile oak (*Quercus petraea*), which is officially Ireland's national tree, and the pedunculate oak (*Quercus robur*). They can be told apart by the position of their fruit: acorns on pedunculate oak grow at the end of hanging stalks (or peduncles), whereas they emerge directly from the twig on sessile oak. The two species have different habitats. Sessile oaks prefer rocky terrain with acid soils and are more commonly found in the west, while pedunculate oaks like heavier lowland soils and are usually found in the midlands and east, although not exclusively. Oaks are the most commonly planted broadleaf in Ireland today.

Birch, hazel, Scots pine, elm and oak were joined by a number of other tree species that filled niches in the understory of the forests and around the forest edges. They included yew, alder, ash, aspen, blackthorn, crab apple, hawthorn, holly, rowan, strawberry tree and whitebeam. Some of these trees went on to dominate large areas in Ireland at different times, but initially they played more of a secondary role.

There are a number of tree species that did not make it to Ireland at this time. Many are important to European forestry today. Many flourished when later introduced to Ireland by humans, so there was nothing about the Irish environment that precluded them. They include beech (*Fagus*), which lords it over the forests of central Europe, and sycamore (*Acer pseudoplatanus*), whose wood is prized for making furniture and musical instruments. Other notable absences are the firs (*Abies*) and spruces (*Picea*), which are the engines of commercial forestry in Ireland today.

Indeed, a low level of biodiversity is a feature of Ireland's flora and fauna. Ireland has 815 native plants, compared to 1,172 in Britain and 3,500 in France. When it comes to trees, Ireland has thirteen native tree genera. In comparison Britain has eighteen, western Europe supports twenty-two and the eastern United States of America boasts fifty-one. This is partly because Ireland is a small landmass with a uniform climate, so the range of habitats is small. But a primary explanation is the history of postglacial migration. As northern Europe froze during the preceding millennia, the forests retreated south, finding refuge in ice-free Spain, Italy and the Balkans. These refugia provided the seed source for the recolonisation of all of Europe, including Ireland, after the end of the last Ice Age. The trees gradually moved north as the ice sheets melted. Some made it to Ireland but others didn't.

The repeated glaciations and Europe's topography also explain why Europe has a lower diversity of tree species than North America. Europe has stretches of sea from east to west – the

Celtic Sea, English Channel and Baltic Sea. Its highest mountain ranges are also oriented in an east–west direction – the Pyrenees, but especially the Alps. These formed natural barriers that made it harder for trees to migrate south as the climate cooled and then to return north when it warmed up. Some trees made the leap, while others fell at the fence. Through multiple glacial cycles, this gradually whittled down the number of trees in northern Europe. In contrast, the main mountain ranges of North America are oriented north–south, so they did not block the smooth migration of tree species in response to changing temperatures. As a result, North America preserved more of its biodiversity.

WE KNOW WHICH TREES made it to Ireland and when. But the exact mechanism by which they arrived has long puzzled scientists. It has tended to pit paleo-ecologists (who study the fossilised remains of plants) against paleo-geologists (who try to reconstruct the geological history of the earth).

Our knowledge of early Irish forests comes from the study of one of the most minute parts of a tree – its pollen. The pollen of each tree species is unique, and can be identified under a microscope. Pollen has been falling on Irish bogs and lake mud for thousands of years. The pollen builds up, layer by layer, preserved in the oxygen-depleted environment. Since the 1940s, scientists have taken core samples from bogs and lakes, separating out each layer, and painstakingly identifying and counting the pollen. In more recent decades, radiocarbon dating has been used to place an age on each layer. Scientists can then tell what trees were present in an area and in what density, at any point in time, peering into an unbroken record that stretches back more than 10,000 years. This analysis has been repeated in more than 400 bogs and lakes around Ireland. It has allowed scientists to identify the trees that were present after the end of the Ice Age. It has also allowed them to create isochrone maps showing the pattern of the migration of tree species across the country.

The results are surprising. Birch was found all across the country in a short period of time, and hazel established itself in the east and south before moving in a north-westerly direction. This indicated that both species probably migrated from Britain across the present-day Irish Sea. But pine, elm and oak were found first on the southern coastline of Ireland and then migrated in a northerly direction over a 500-year period. British pollen records show that these species arrived in Britain around the same time. Therefore pine, elm and oak could not have come to Britain first and then crossed the Irish Sea; they were not British hand-me-downs. Instead they appear to have migrated from Spain along the Atlantic French coast and then directly to Ireland across the Celtic Sea. This is supported by DNA analysis, which shows that all Irish oaks are derived from Iberian ancestors, whereas British oaks share Iberian, Italian and Balkan lineages. There are fifteen plant species – and some insects too – found in Ireland, but not in Britain, whose nearest relatives are found in Spain. A famous example is the strawberry tree (*Arbutus unedo*), which grows especially well in County Kerry.

The evidence suggests that trees came to Ireland both from Britain to the east and directly from the European mainland to the south. How did they get here? The gap between Torr Head in Antrim and the Mull of Kintyre in Scotland is just 20 kilometres, and on a fine day you can see the peaks of Snowdonia in Wales from the top of the Dublin Mountains. But the distance between the south coast of Ireland and the French province of Brittany is a daunting 450 kilometres. How could trees have crossed this expanse?

One theory is that the sea was not, in fact, there at all. Instead, trees migrated across land bridges that connected Ireland with Britain and continental Europe. It is difficult now to picture these land bridges. There are troughs in the seas around Ireland that are at least 100 metres deep. But sea levels were much lower at the end of the last Ice Age, for two reasons. First, huge volumes of water had evaporated from the oceans, fallen as snow on land, and become locked up in giant ice sheets. As a result, there was a lot less water in the oceans. Second, the Ice Age, and its ending, did funny things to the earth's topography. The weight of kilometres of ice pressed the land down, and, once it was removed, the land rebounded. The seabed got higher. There may even have been a temporary 'forebulge', caused by rippling of the crust, which travelled from the south-west to the north-east in front of the receding ice sheets. If the rebound of the land as the weight of ice was lifted temporarily outpaced the rise of the sea as the glaciers melted, this could have been enough to create a ridge of dry land around 16,000 to 9,000 years ago, stretching across the gaps that currently separate Ireland from Britain and France.

The existence of temporary land bridges would explain why Ireland has certain trees and not others. The swift and hardy made it to Ireland from their southern refugia while the bridges existed. But other species didn't make it. For example, the lime tree crossed to Britain from continental Europe around 8,000 years ago and had made its way to the Welsh coast 7,000 years ago, but it was assumed that it did not make it to Ireland. Similarly, the slow-moving beech, which could only migrate less than half a kilometre each year, arrived in England much later, around 3,000 years ago, and found its way to the Welsh coast 1,000 years ago, but also failed to cross the Irish Sea. The drawbridge had been pulled up. Ireland had become an island, protected from further migrations. This allows us to draw a clear distinction between native tree species that arrived through 'natural' processes soon after the end of the Ice Age, and species that were imported by humans later on.

The land bridge theory has neat explanatory powers. It has been most strongly supported by paleo-ecologists who study pollen records. But the geologists who study rocks, seabeds and ice sheets have always had their reservations. And, more recently, the academic pendulum has swung against it. Two academics from Trinity College Dublin, Robin Edwards and Anthony Brooks, who modelled the response of the earth to the loading and unloading of glacial ice, flatly reject the idea. Their research indicates that Ireland had been separated from Britain and the rest of Europe by 16,000 years ago, long before the first trees arrived. 'No support is found for the idea that a Holocene land bridge was instrumental in the migration of temperate flora and fauna into Ireland,' the researchers baldly state. Just in case the paleo-ecologists weren't paying attention, they titled their paper 'The island of Ireland: drowning the myth of an Irish land-bridge?'

Instead of a continuous land bridge it seems that tree seeds were carried to Ireland by water, wind or birds – what scientists call 'vectors'. Every year, for example, viable coconuts and other seeds from the Caribbean islands wash up on the shores of County Kerry. There are even seeds from West Africa that have drifted across the Atlantic to the Caribbean before being carried by the Gulf Stream to Ireland. Storms can pick up light seeds and deposit them far from their source. Perhaps most importantly, Ireland is on the migratory path of birds travelling between southern Europe and the Arctic – seeds can hitch a feathered ride. Lower sea levels and the rebounding of the earth's crust in the postglacial period may have temporarily shortened the distance between the Irish coastline and the rest of Europe, or created a chain of islands, making it easier for vectors to do their work. Whatever way this process operated, Ireland was not as isolated as we once thought.

WE DON'T KNOW EXACTLY how trees migrated to Ireland after the ice sheets disappeared, but there was certainly some transmission mechanism that allowed for the rapid population of Ireland with trees. The accepted wisdom is that sometime around 9,000 years ago all the trees that are now regarded as native to Ireland were in place. Over the next two thousand years, the forests expanded and developed until almost the entire country was covered in trees. There was a mosaic of different woodland types, depending on soil type and climate. The lowlands and better soils supported dense mixed broadleaf forests, chiefly of oak and elm. Scots pine dominated poorer sites, especially in the west. Hazel shrubs occupied the understory beneath these trees in all parts of the country. Alder and willow could be found on marshy land near lakes and rivers. A number of other species filled niches throughout the forest – they included birch, yew, aspen, blackthorn, crab apple, hawthorn, holly, juniper, rowan, strawberry tree and whitebeam. This range of trees – small in global terms – provided the palette for the evolution of the Irish landscape over thousands of years until humans started to introduce 'exotic' species from other areas. When people talk about 'native woodland' today, they mean these species and not others.

Yet it turns out that 'native' is a slippery term. It is a two-dimensional concept, defined not just by place but also by time. The Holocene period (roughly the last 12,000 years) is a snapshot in time. If we look further back, we can find a much broader range of trees in Ireland.

The Pleistocene, which lasted from about 2 million years ago to 12,000 years ago, saw at least eight freeze-thaw cycles. Temperatures dropped and ice sheets stretched their fingers forward, only to withdraw as the planet warmed up. These mild interglacial periods, when temperatures were sometimes a few degrees warmer than now, lasted for thousands of years. Forests receded in the face of advancing ice and then migrated back, ebbing and flowing like tides on a beach. During these interglacial periods, Ireland had a much broader range of trees (as well as animals such as the woolly mammoth, hyena and brown bear). For example, all the twenty-two tree genera native to Europe were present in Ireland 240,000 years ago. They included beech, lime and hornbeam as well as firs and spruces. These were all 'native' to Ireland at one time. But around one-third of these species didn't immediately make it back after the last postglacial migration. If we zoom our lens back further, the list of Irish trees starts to look even more exotic. This is because the landmass we now call Ireland has been on the move for millions of years. Its passport is well stamped. One part of Ireland started off near the South Pole. Carried by the movement of tectonic plates it passed through the equator, when the climate would have been tropical, before moving through the mid-latitudes and ending up in its current position. All sorts of trees – tropical, subtropical, temperate and arctic – grew on the land as this trans-hemispheric migration played out. They show up in fossils across Ireland. For example, mining operations have uncovered fossilised tree stumps identifiable as cypress (*Cupressus*) and monkey-puzzle (*Araucaria*).

Recent research is also challenging long-standing assumptions about which trees should be placed in the 'native' category and which should be regarded as later human introductions. Scots pine provides one fascinating example. As we shall see, over many thousands of years the range of this tree gradually shrank because of a changing climate and human impact. It was long believed that the tree had become extinct by around AD 400. All Scots pine trees in Ireland today, it was thought, owed their origin to seedlings brought from Scotland and planted by landowners from the eighteenth century onwards. This placed Scots pine in an ambiguous position as part-native, part-import. For example, the current Native Woodland Scheme provides grants to plant it, whereas the Irish Peatland Conservation Council considers it an invasive, non-native species.

In 2009 a PhD student in the Department of Botany at Trinity College Dublin published a paper that threw this convention on its head. Jenni Roche had been studying Scots pine forests in Ireland when she visited Rockforest Wood, a small site 10 kilometres north-east of Corofin in the Burren, County Clare. It looked more ancient than anything else she had seen, and there was no record of it having been planted. It prompted her to dig up mud samples from a nearby lake and to analyse the pollen, fossilised needles and wood fragments buried within. Much to her surprise she found that there was an unbroken record of Scots pine stretching back over the last 2,000 years. This proved that Scots pine did not become extinct in Ireland but had survived in relict woodlands, such as this example in the karst limestone landscape of east Clare. Further studies in the area confirmed these findings.

While Scots pine may never have died out, other research indicates that one species long accepted as native may actually have been introduced by humans much later on. The strawberry tree occupies a special place in Irish botanical lore, as it is one of a group of 'Lusitanian' species that originated on the Iberian Peninsula and occur in western Ireland but not at all in Britain. It is today considered native to counties Kerry, Cork and Sligo. It is first found in the pollen record around 4,000 years ago, long after the initial tree migration to Ireland. Micheline Sheehy Skeffington and Nick Scott from NUI Galway argue that it was introduced by people from northern Spain who brought copper mining to the south-west of Ireland around that time. It could have been introduced because its berries can be used to produce alcohol – they are distilled for brandy in Mediterranean countries today. Indeed, the latest research indicates that most of the 'Lusitanian' species, such as the Kerry slug, were introduced by humans over the last few thousand years.

A finer analysis of pollen has also thrown doubt on when other tree species arrived in Ireland. It turns out that scientists often find species that should not be present when studying pollen records, but they usually dismiss this as being due to contamination. In 2017 Susann Stolze and Thomas Monecke of the University of Colorado carried out a detailed study of 500 Irish pollen records and reached a different conclusion. Through a process of elimination, they argue that at least two species – lime and hornbeam – were established in Ireland around 8,000 years ago and should be considered as native trees. They also found that other non-native species, such as sycamore and beech, had been present in Ireland for much longer than thought, and had become naturalised. Their conclusions are not yet widely accepted, but they cast further doubt on the notion that there was a distinct cut-off after which no new tree species migrated to Ireland. Instead there may have been a gradual seeping of species from the rest of Europe.

Two things stand out from this broader view. First, the definition of 'native' is a little arbitrary. It is supposed to refer to those trees that managed to scramble back to Ireland after the last Ice Age through 'natural' processes, before humans started to influence the environment. But if we pick a different point in time, we can find a broader range of trees in Ireland. And it is not clear when certain species arrived, if others died out and whether or not humans were involved in their introduction. Environmentalists sometimes make a fetish of 'native-ness' when discussing current forest policy. Perhaps there is a better way of valuing a tree, rather than the simple test of having made it back to Ireland in the few thousand years after the end of the last cold snap. This is a mere blip in geological time. We can expect that tree species would continue to migrate and to find ecological niches to put down roots. Looked at from this perspective, humans are just another vector in this process, not so different from birds or ocean currents.

Second, climate change is nothing new. The climate is *always* changing, in response to solar cycles and the amount of carbon dioxide in the atmosphere. Ireland has sweltered through warmer epochs than today, and has shivered through intense Ice Ages. The country's flora and fauna have adapted to these changing conditions. There has been no constant. Eventually it becomes too warm, too cold, too dry or too wet for particular trees. Then the trees migrate to other climes and the gap is filled by better-adapted species. It looks like we are about to embark on a period of rapidly warming temperatures, this time caused by human activity. More change is coming, and Ireland's trees will have to adapt, as they have always done.

THE CLIMATE CONTINUED TO change after the initial reforestation of Ireland during the early Holocene. As we shall see, this would alter the composition of Irish forests, creating winners and losers. But the flora of Ireland gradually came under the influence of a new type of fauna – humans. These new migrants followed the advancing forests from their own refugia in southern Europe, most likely arriving in Ireland via boat. As well as bringing new seeds and animals, they brought fire, axes and an endless ability to innovate. Human impact and climate change intertwined in subtle ways to reshape Irish forests for thousands of years. But human activity later became the dominant force, unleashing a level of arboreal destruction that had not occurred since the arrival of the ice sheets. The Irish landscape would never be the same.

EARLY HUMANS

IN 2016 A DUSTY BOX of bones that had been sitting in the basement of the National Museum of Ireland for more than 100 years yielded up a great secret. It was part of a hoard of bones that had been discovered in a cave near Ennis in County Clare during an excavation in 1903. More than a century later two researchers – Marion O'Dowd and Ruth Carden – decided to reopen the boxes and take another look. They were sorting through the bones when they came across the patella, or knee bone, of a brown bear. It had seven cut marks, indicating that it had ben butchered by a human hand with a flint blade. When they sent the bone for radiocarbon dating, the results showed that the bear had been killed and butchered 12,500 years ago. Doctors O'Dowd and Carden had just found the earliest evidence of humans in Ireland.

This was such an exciting discovery because it revealed that people were in Ireland during the Palaeolithic period, 2,500 years earlier than previously known. It means that humans were around before trees migrated to Ireland after the last Ice Age. These first Irishmen and Irishwomen would have lived in an open landscape of grassland and shrubs, when the first birch and juniper trees were establishing themselves. They would have hunted the Irish giant deer and red deer that roamed at that time. We do not know if they survived the cold snap that occurred around 12,000 years ago. They may have died out, or left Ireland when the climate worsened. Or there may have been a constant human presence, which would mean that our ancestors witnessed the arrival of the first pine, oak and elm and the growth of mighty forests.

Apart from a box of bones in the National Museum we have very little evidence of Palaeolithic humans in Ireland. The waves of Mesolithic people who began to turn up in Ireland around 9,000 to 10,000 years ago (or 7000 to 8000 BC) left more traces. They arrived by boat, most likely from Britain, as similar tools and settlements have been found there. They are known for their flint blades, which were used as cutting tools on their own or inserted into the end of wooden poles to make spears. They mostly lived on the coasts or beside lakes or rivers. The Mesolithic people of Ireland had very little impact on the forests. Each settlement had a catchment area that was exploited for raw materials. The small communities cut trees for firewood, tools and containers. They ventured into the forest to collect plants, fruits and nuts, especially hazelnuts, and perhaps to hunt for animals such as wild boar. Fish and shellfish were a big part of their diet. For example, excavations at a Mesolithic settlement at Mount Sandel in County Derry show that the inhabitants dined on eels, salmon, hazelnuts, berries and wild boar. But agriculture and domestication of animals had yet to appear, and there was no clearing of woodlands on any scale. Overall, Mesolithic communities left a light footprint. Their total population was probably no more than five or ten thousand at its peak.

Irish woodlands did undergo changes at this time, but it was due to climatic factors. Around 7,000 years ago, the island came more strongly under an Atlantic influence, and rainfall increased. The tree that benefited most was alder (*Alnus glutinosa*). Alder had arrived early in Ireland, but was held in check until the shift to wetter conditions, from which time it starts to appear more and more in pollen records. Alder loves damp soils and is often found growing beside lakes and along riverbanks. It can grow into a substantial tree in the right conditions. In the past, it was used to make clogs, and it was popular in the furniture trade, where it was known as 'Irish mahogany'. Because it does not decay in water, alder was also used to make sluice gates and other structures along rivers and canals. Alder has proved a canny survivor in Ireland's damp woodlands, and is now one of the most widely distributed trees in the country.

Ireland was densely forested at this time. In fact, the extent of forest cover increased during the Mesolithic period. Ireland achieved its highest proportion of woodland cover, what you could call 'Peak Forest', around 6,000–7,000 years ago. The climate was stable, with a temperature a little warmer than today, especially in the summers. Scientists nickname it the 'Climatic Optimum'. The soils had developed sufficiently since the scouring of the Ice Age to support a wide range of trees. And these trees had been in situ long enough to fill every niche and to establish a complex, inter-dependent community under an umbrella of oak, elm and Scots pine. Humans occupied their own niche within this ecosystem and were too small in number to disrupt it. This early woodland stretched from shore to shore, across almost all the lowlands and extending up the sides of most mountains. The west of Ireland, now treeless and covered by a blanket of bog, was mostly forested. The only places without trees were areas of exposed limestone rock, lakes and wetlands, unstable screes and the most exposed mountaintops. In all, around 85 per cent of the country was covered in forest. The situation in Ireland mirrored that across Europe, which reached a 'forest maximum' between 8,000 and 6,000 years ago.

The primeval Irish oak forest was a wild and dark place. Giant trees, metres across, towered over the canopy. The understorey was a tangle of shade-tolerant trees and shrubs. There was plenty of deadwood rotting on the floor, supporting a huge diversity of fungi and insects. The forest was so dense that hardly any grasses or flowers lived under it. The closest parallels in Europe today are the Białowieża Forest in Poland, although it contains different species to Ireland and has been subject to human intervention, or areas within French oak forests that are kept as nature reserves, with no paths, no human visitors and no felling or tidying of the forest. As Jonathan Pilcher and Valerie Hall write in their book *Flora Hibernica*, 'they are awesome and frightening places'. This is the *selva oscura*, the dark wood, that the Italian poet Dante found himself in at the beginning of his epic poem the *Inferno*.

AROUND 6,000 YEARS AGO a new wave of migrants landed on Irish shores – Neolithic farmers. The new settlers carried crops such as wheat, domesticated animals and polished stone axes. These new technologies had been introduced by people from the Near East, who gradually moved through Europe displacing the existing populations, and to some extent breeding with them, before reaching Ireland. Their arrival marked the shift from a society based on hunting and gathering to one based on agriculture. They were hungry for open land on which to grow crops and raise cattle and sheep. They built impressive megalithic tombs, such as the enormous burial mounds at Newgrange and Knowth in County Meath, or the dolmens and wedge tombs scattered throughout the island. Unlike their Mesolithic predecessors, their environmental footprint was large.

We know a lot more about the lives of early humans in Ireland now thanks to the building mania of the Celtic Tiger era. This was a boon to archaeologists. Salvage excavations were often required by law before new housing estates were built or new roads blasted through the countryside. By 2006 more than 2,000 archaeological digs were taking place per year, compared to just fifty in 1970. Some extraordinary linear digs preceded the new motorways radiating out from Dublin: for example, fifty-six archaeological excavations were carried out across a 64-kilometre route in advance of the Kinnegad-to-Athlone M6 road scheme. These digs have thrown up a trove of information on prehistoric Ireland and given birth to hundreds of academic papers.

Researchers use a range of techniques to understand how people interacted with the landscape thousands of years ago. Pollen-grain analysis is still a mainstay, as it helps to map the presence of different plants and forest types. Charcoal analysis tells us what type of wood was burnt for cooking, heating and, later, for metalworking. There is one other technique that comes into its own when reading the story of trees – tree-ring analysis, or dendrochronology. Tree rings vary in diameter each year because weather conditions cause trees to grow slower or faster. Each sequence of tree rings is unique to its time, as it reflects a particular sequence of good or bad years. By lining up old building timbers and preserved wood pulled from bogs, researchers at Queen's University Belfast have created an unbroken chronology of oak tree rings that now stretches back 7,308 years. They can take any oak sample in Ireland, line it up against their tree-ring chronology and tell you exactly when that tree was growing and when it died. In effect, they can read tree rings like a barcode.

Some of these 'barcodes' are found on the stumps of trees felled by Neolithic settlers. We tend to criticise people in developing countries for clearing rainforests for agriculture or timber, yet our Neolithic ancestors did just the same. They cleared forests to sow crops such as wheat and barley. They cleared bigger areas to graze cattle and sheep, setting in motion the long transformation towards an island dominated by pasture. They could be remarkably effective at clearing primary forest with the tools at their disposal. Trees were ring-barked (removing a strip of bark around the entire circumference of the trunk) and left to die before fire was used to clean up undergrowth. Polished stone axes with long wooden handles, swung from the elbow with short, sharp cuts, could fell even the largest trees. Experiments in Denmark showed how effective the Neolithic stone axe could be. Using an axe head that had not been sharpened for 4,000 years, three men managed to fell 100 trees and clear 600 square yards of silver birch forest in four hours.

An extraordinary example of the clearance of trees from the land can be found at Céide Fields on the northern Mayo coast. This stretch of heathery bog on the edge of a cliff beside the Atlantic Ocean looks unprepossessing at first, but underneath is what has been called the most extensive Stone Age monument in the world, older than the Pyramids. It is certainly the oldest example of a Neolithic field system in all of Europe. Céide Fields is a regular system of stone-wall-bounded fields and megalithic tombs that stretches over several square kilometres. Around

6,000 years ago this area was covered in forest dominated by pine, along with hazel. A layer of high charcoal content indicates that this woodland was set on fire and cleared around 5,750 years ago. A Stone Age community developed here, practising intensive farming for around 500 years. Most of the land was converted to grassland to raise cattle and sheep, but cereals were also grown. Similar forest clearances, known as 'landnam', have been found throughout Ireland. They show up in the pollen record: the quantity of tree pollen decreases, new pollen from cereal crops shows up and the amount of grass pollen increases, indicating a shift from woodland to pasture.

The arrival of Neolithic farmers coincided with other changes to Irish woodland. One of the most abrupt was the dramatic decline in elm. Elm had played an important role as a dominant canopy tree alongside oaks, especially in the midlands. Around 5,500 years ago it suddenly drops out of the pollen record. And not just in Ireland, but all across Europe. In the early years of paleo-botanical research, it was thought that the decline of elm was brought about by Neolithic people cutting off the young branches to feed cattle, as young elm foliage is both sweet and nutritious. But the experience of Dutch elm disease in the 1960s and 1970s changed perceptions. It is now thought that a similar fungal disease struck elm during the Neolithic period. Elm would make sporadic comebacks in Ireland and still survives in occasional hedgerows today up to an immature stage, but it never reclaimed its starring role alongside oaks in the early story of Irish forestry.

The other major change was the growth of peat bogs, an iconic – and often melancholy – feature of the Irish landscape that would have an important influence on Irish forestry to the present day. Bogs form when soils are saturated by constant rain and do not have a chance to dry out. The rain washes minerals from the topsoil and forms an impenetrable iron pan below, which causes permanent waterlogging. Mosses, heather and sedges come to dominate. In this oxygen-starved and acidic environment, dead plants do not decay and instead accumulate, causing the peat layer to grow higher and higher, sometimes up to 12 metres thick. There are two types of bog in Ireland: blanket bogs, which cover mountains and large areas of the west; and raised bogs, which grew out of lake basins in lower parts of the midlands. Today, some 16 per cent of Ireland is covered in peat bog, which is the third-highest proportion of any country in Europe (behind Finland and Estonia).

Bogs began to form around 4,500 years ago. They expanded outwards from depressions and encroached up and over hillsides, eventually covering thousands of square kilometres across Ireland. For example, the Bog of Allen in the midlands covered nearly 1,000 square kilometres at its peak. Along the way, bogs swallowed up Irish woodlands. Roots died in the waterlogged soils, seedlings could not regenerate. As blanket bog developed, the west of Ireland and the uplands lost their woodlands and became open landscapes, cloaked in a gentle brown mantle of heather and moss, creating the classic Connemara landscape we know today.

The biggest casualty was Scots pine, which had been the dominant canopy tree in the west of Ireland and on the mountains. Initially, Scots pine and bogs found a way to co-exist. There were periods when the peat soils were dry enough to allow trees to grow on them. Some of these big pine woodlands survived for 1,000 years. But eventually the wetness was too much and the pines died off. The decline began around 4,500 years ago and then accelerated 2,000 years later. The remains of these ancient pine forests were discovered in the modern era by puzzled turf-cutters, who dug up blackened stumps and logs that had been preserved in the wet, acidic soil. This 'bog wood' is a favourite of contemporary sculptors because of its unique semi-fossilised texture and the story it has to tell. One area where Scots pine clung on for longer was the Burren region of County Clare – home to Rockforest Wood – because the cracked limestone pavement and free-draining soils prevented bog formation.

We can see this process playing out on the Céide Fields as well. After 500 years of farming, the land was abandoned around 5,200 years ago. This change happened quickly, within about fifty years, or two generations. Over time, blanket bog formed, covering the area in a thick layer of peat. This is why the Neolithic field system is so well preserved. It was only discovered in the 1970s when turf-cutters hit the ancient stone walls while digging out the peat. Afterwards, archaeologists spent years probing the bog with metal rods to map out all the walls.

The discovery of the buried Céide Fields led many to wonder what role early humans played in the shift from woodlands to open peat bogs. One of the tragedies of the destruction of modern-day rainforests is that it can permanently degrade the land. The land produces a few good crops, but the soil is soon exhausted and the farmers are forced to move on to the next patch of forest. Did something similar happen in Irish prehistory? Did humans cause the spread of bogs by clearing the forests that kept the ecosystem in balance?

There are some grounds for believing this. Trees improve soil drainage and pump water back into the atmosphere through transpiration. It is easy to see how the clearing of trees could disrupt the hydrological cycle and cause waterlogging of soils. Charcoal from clearance fires is commonly found in the lower layers of peat bogs. Fire destroys plant communities, and ash from fires can clog the drainage pores of soil, again contributing to waterlogging. Once trees were cleared in the wind-pummelled west, they found it very difficult to re-establish themselves, as regeneration was dependent on mutual shelter. And the land would have continued to degrade through overgrazing by domesticated sheep and cattle. According to Jonathan Pilcher and Valerie Hall, authors of *Flora Hibernica*, 'it now seems highly likely that most of Ireland's blanket bog owes its origin directly to forest clearance by people'. If this is the case, perhaps bogs are not such a 'natural' feature of Ireland after all? Perhaps they are a consequence of human deforestation?

If we take another look at Céide Fields, we see that the story is not so simple. After the field system was abandoned 5,200 years ago, it was not immediately swallowed up by blanket bog. In fact, the land returned to full woodland cover, dominated by pine and birch. It stayed this way for hundreds of years. Blanket bog only began to form around 4,500 years ago, eventually overwhelming the pine, which sank into the bog, leaving the stumps for us to peer at today. There are plenty of other examples of blanket bog expanding onto untouched forested land, rather than only on cleared agricultural land. And blanket bog formed across many parts of Europe at this time, even where human influence was negligible.

We seem to gain a dark satisfaction from believing that the actions of people determined the shape of our landscape, and its most degraded features, thousands of years later. It appeals to our homocentric view – we act on the world, rather than the world acting on us. It also taps into a deep-rooted shame, at least in western civilisation, around 'the fall of man' and the original sins that led to our expulsion from a sylvan paradise. Yet, according to the latest academic research, blaming human action for the formation of bogs may be overdone. A changing climate probably played a bigger role, as summer temperatures fell and there was a shift to wetter and cooler conditions. Forest clearance may have accelerated the process, but nature was more than capable of reshaping the landscape on its own. As Professor Fraser J.G. Mitchell of Trinity College Dublin points out, there was no such thing as a 'climax forest community', static through time. The history of Irish forests is a 'dynamic one', driven by the interplay between a changing climate and human activity.

The level of human activity in Ireland was also dynamic and changing during the Neolithic period (6,000 to 4,250 years ago). Forest clearance was patchy and interrupted, rather than a one-way process playing out across the whole island. There was an abrupt transition to agriculture around 5,750 years ago, especially in the west and north-east. But, as with Céide Fields, areas that were cleared for agriculture often reverted to forest after a few hundred years. There seems to have been a lull in farming activity and a fall in population during the later Neolithic period, around 5,300 years ago. This may have been linked to a worsening climate, as cooler, wetter conditions made farming more difficult, or it could have been triggered by social and economic decline. One academic paper refers to the 'the boom and bust of early farming in Ireland'. This mirrors the pattern in Britain, where the high levels of human impact reached in the earlier Neolithic were also followed by a decline in farming activity and the regeneration of woodland.

Human impact on forests during the Neolithic period was also very different from region to region. Forests were cleared most rapidly in the west of the country, which was the original focus for settlement. This could have been because of easier access via the sea and waterways, or because the more open pine woodlands were easier to clear. Woodland cover in the west declined from around 80 per cent to 70 per cent during the Neolithic. In contrast, the midlands maintained full forest cover throughout the period and were largely untouched by people. In the east of the country, woodland cover was around 80–95 per cent over the course of the Neolithic, with more settlements starting to appear towards the end of the period.

Overall, there was a shift in farming activity from the west to the east over these 1,750 years. This is consistent with a change in climate, as the west would have suffered most from wetter, cooler conditions. It is surprising to discover that Neolithic settlers picked the northern coast of Mayo for home, when this region is almost depopulated now. There is a reason that Mayo is home to Ireland's first International Dark Sky Park, showcasing some of the darkest, most unpolluted skies in the world. The bright lights are now further east.

THE LATE-NEOLITHIC LULL was followed by another burst of forest clearance as new migrants ushered in the Bronze Age around 4,250 years ago. According to the latest genetic research, these people can be traced back to herders who originated in the Pontic steppe, which stretches from the northern shore of the Black Sea to the Ural Mountains. They migrated across Europe and eventually made it all the way to Ireland. They brought new metalworking techniques, centred around bronze (a mixture of copper and tin) and gold. They had a highly organised society, holding communal gatherings at sacred places on the landscape and burying hoards of weapons, jewellery and other objects as offerings to the gods.

We call this the 'Bronze Age' because spectacular metal objects survived to be found and proudly displayed in museums. But this was just as much a 'Wood Age'. Wood was the most widely used material in everyday life. People lived in houses made of timber and wattle (rows of stakes interlaced with twigs or branches). They cooked by burning wood at *fulachtí fiadha*, large pits sunk into the earth where heated stones were used to boil water – over 4,500 examples have been found around the country. Wood was selected for its different traits, such as durability and elasticity. Hazel and oak were most commonly used for buildings and firewood. Vessels and tubs required for domestic use were frequently made of ash, alder and oak. Yew, ash and hazel were used to make the handles of daggers, knives and razors. Axe handles were often made from alder.

Wood was also how people got around. Extensive wooden trackways were built across bogs. The people who came to Ireland must have come in boats large enough to carry domesticated animals, which means they were skilled in working timber to build boats. The most impressive wooden artefacts to date from this period are dugout canoes. One example is the Addergoole canoe, which is now in the National Museum of Ireland. It was found in a bog in County Galway in 1901 by Patrick Coen as he was cutting turf. Dating to 2200 BC, the boat is a metre wide and 14 metres long, which is much longer than a Dublin bus today. It was carved from the trunk of a single oak tree that was at least two metres wide, which shows how big the old-growth trees were at this time. It took more than a month to move the vessel from Lurgan to the National Museum in Dublin by train and specially linked horse-drawn carts. The canoe is so big that it was probably used for ceremonial purposes by the ruling elite. According to Eamonn Kelly of the National Museum, 'it has a touch of the late-Neolithic Porsche about it'.

The Bronze Age people plundered the forests to provide materials for everyday life. They also cleared a lot more forest for agriculture. They were enabled by a new technology, the bronze axe, which was more effective than stone when it came to felling large trees. For the first time, the great old oak woodlands of the midlands began to fall. Bronze Age settlers pushed into Kerry and Cork, as this was where they found copper to mine. And the number of human settlements in the east of the country increased. The population of Ireland was much higher than the previous Neolithic period, which increased demand for land and resources everywhere. It is estimated that woodland cover had been reduced to around 50 per cent of the total land area by the end of the Bronze Age (circa 500 BC).

Human action also began to have a noticeable impact on the composition of the forests and the distribution of species within the forests. One tree that suffered was yew (*Taxus baccata*). Yew is one of only three conifers native to Ireland (alongside Scots pine and juniper). It is easily recognised by its pointed, flat, dark-green evergreen needles. Yews can live for more than 1,000 years and have the longest lifespan of any European tree. For example, Brian Boru, who was slain by Vikings in 1014, was said to have died under a yew that could be found growing in a garden in Clontarf, County Dublin until it toppled over in 1993. To early Christians, a yew was a symbol of eternity and immortality because of its longevity, and because it held its leaves through winter.

Yew expanded rapidly in Ireland around 4,850 years ago, especially in the west and south-west. It is resistant to wind and, for a time, was able to grow on shallow peat soils alongside Scots pine. It formed mighty specimens that grew to an impressive height, quite unlike the small stunted trees that mostly survive today. Beautifully preserved fossilised yew trunks often turn up in bogs alongside pine and oak. But its expansion was short-lived, often only lasting a century in some areas, before declining for good.

Although the growth of bogs probably played a role, human impact was mostly responsible for the demise of this tree. Yew woodlands were cleared to make way for farming and domesticated animals. The yew tree is very sensitive to grazing, so it found it hard to regenerate. Its timber was also favoured for tools. One example is a beautifully carved mallet of yew found in Inchagreenoge in County Limerick as part of excavations before the construction of a gas pipeline in the early 2000s. The head was shaped from the trunk of a tree, with a branch as its handle. Archaeologists believe it may have been accidentally lost by its Bronze Age owner, as it was in excellent condition. Yew was used for weapons and tools across Europe at this time. A very similar yew mallet has been recovered from peat in the Somerset Levels in England. Further afield, the Tyrolean Iceman ('Ötzi'), whose frozen body was found in a glacier on the Austrian/Italian border, had an arrow and an axe handle made from yew.

There are not many yew woodlands left in Ireland. The most famous is Reenadinna Wood in Killarney National Park, which has been there for more than 3,000 years. It is an eerie place, with lichen dripping from the ancient branches and moss carpeting the floor, bathed in Atlantic moisture. Yew is relatively uncommon in the wild. It is most often found in churchyards and on early Christian sites. Its legacy survives in the names of townlands, at least 160 of which are derived from this tree. One example is Terenure or 'Tír an Iúir' in Dublin, which means 'Territory of the Yew'.

Other trees benefited from the actions of Neolithic and Bronze Age farmers in Ireland. One such tree was the ash (*Fraxinus excelsior*), a large deciduous tree that grows up to 40 metres. Its flowers are very dark, almost black, and may be seen before the leaves develop – ash is one of the last trees to come into leaf, and is one of the first to lose its leaves in autumn. Ash timber is hard-wearing and can be put to many uses. It is flexible and resistant to splitting, which is why it is used to make hurleys as well as snooker cues, hockey sticks and oars.

Ash had been around since the development of the first postglacial woodlands in Ireland but only as a minor species. It became much more common in the pollen record around 6,000 years ago. This coincided with the arrival of the first farmers. Ash appeared in the wake of woodland disturbance, especially in limestone areas in the midlands and the Burren. It recolonised areas that had been cleared for farming and then abandoned, or grew around the edges of fields. In a way, ash was the new birch. It became widely used in the Bronze Age, often appearing in charcoal finds. Ash proved a hardy survivor for thousands of years and is possibly the most common hedgerow tree in Ireland today. However, as we will explore later, it is now threatened by ash dieback, a new fungal disease that is creeping through the island and threatening to do to ash what Dutch elm disease did to the elm.

People were also responsible for introducing new tree species to Ireland during this period. As we saw in the last chapter, the strawberry tree may have been introduced by copper-mining people from northern Spain around 4,000 years ago. Beech (*Fagus*) was one of the most common species in mainland Europe at that time, and beechnuts were an important food. Its use in Ireland is first detected during the Neolithic period. It was probably a deliberate introduction by early farming groups. It has now become naturalised and occurs frequently in woods and hedges. Walnut (*Juglans*) also appears in pollen records around this time. It is known for its highly nutritious fruits and the medicinal properties of its bark and leaves. Sycamore (*Acer*), or sycamore maple in the United States, was another introduction by Neolithic people, and had become widespread by the Bronze Age. It is now one of the most common hedgerow trees in Ireland. Beech, walnut and sycamore had been present in Ireland before the last Ice Age. Early humans were an effective vector for completing the migration back to Ireland of species that had not been able to make it on their own.

By the end of the Bronze Age there was a greater diversity of trees in Ireland, but much less forest overall. Forest cover, which may have reached 85 per cent during the Mesolithic period, had declined to around 70 per cent by the end of the Neolithic, and to around 50 per cent during the Bronze Age. There were large areas of open landscape, used for cereal farming and grazing domesticated animals, in every part of the country. This is evidenced in pollen analyses by a rise in ribwort (*Plantago lanceolata*), a common weed of cultivated land. There was no single clearance event, but rather a gradual process of opening of the landscape. Forests were cleared all across Europe at this time, starting around 6,000 years ago and then accelerating 4,000 years ago. But Britain and Ireland saw the most rapid forest loss, early on. Although on the fringes of the Continent, Ireland was at the vanguard of this particular marker of civilisation.

Yet, as before, deforestation was not a linear process. It proceeded in fits and starts. Land that was cleared for agriculture could be abandoned for hundreds of years, allowing woodland to regenerate. There were large areas of what are called 'secondary forests'. There were cycles of forest-clearing and regeneration. Toward the end of the Bronze Age there was another lull in human activity. Farmland was abandoned, and the population stagnated or declined. More weapons (swords, spears, rapiers, knives, dirks and halberds) are found in buried hoards, which may indicate a rise in warfare between communities. Researchers studying ice cores in Greenland believe there was a worsening of the climate around 1000 BC, leading to colder winters, wetter summers and more frequent Atlantic storms. The countryside may actually have become more heavily forested between 1000 BC and 500 BC as the Irish population declined.

Over 10,000 years environmental factors and human activity intertwined to shape Irish woodlands. Climate change caused some tree species to flourish – such as alder and, briefly, yew – while leading to the growth of bogs that would swallow up huge areas of Scots pine forest. The arrival of an arboreal pandemic from the east is the most likely explanation for the decline of elm. Human activity then became the most powerful influence on the evolution of Irish woodlands. Humans affected the composition of forests. They selected out certain species favoured for timber, such as yew; they acted as a vector for the introduction of species like beech and sycamore that had not yet migrated back to Ireland following the end of the Ice Age; and they created the conditions for ash to play a more dominant role. But, above all, humans cleared forests for agriculture, resulting in a landscape that was at least half open by the end of this period. They set in motion a process of deforestation that would continue, in fits and starts, until the twentieth century.

HENCEFORTH, HUMAN ACTION, rather than climate or pathogens, would be the most powerful influence on Irish forests. This work was carried on in the next phase in Irish history – the Iron Age. Iron replaced bronze in tools and weapons as a technological revolution spread from continental Europe. New forms of art and settlement appeared. The Irish language, a Celtic language that survives today, became the dominant tongue. An oral culture passed down by poets was eventually captured by scribes in written form, giving us our first glimpse into the place of trees in the minds and culture of the Irish.

CELTIC IRELAND

THE EARLIEST WRITTEN REFERENCE to Celts is from about 500 BC, when a Greek geographer, Hecataeus of Miletus, wrote about *Keltoi* in his work. Roman literature contains many descriptions of the Celtic peoples who battled with – and in some cases were subjugated by – the Roman empire in central and western Europe. Celts were seen as having a special affinity with forests, using them as shelter from Roman legions and as sites of worship. Tacitus, writing about the Roman campaign on Anglesey off the coast of Wales in AD 60, described how imperial troops demolished a Celtic woodland grove devoted to 'barbarous superstitions'. 'It was their religion,' he explained, 'to drench their altars in the blood of prisoners and consult their gods by means of human entrails.' The image of long-haired druids in flowing robes clutching branches and performing rituals under oak trees derives from these classical descriptions.

How applicable are these depictions to the people of Ireland at this time? Setting aside the obvious prejudices of classical writers, it depends on the extent to which there was a common Celtic culture stretching across Europe. There certainly was a family of Celtic languages, common technologies based on ironworking and shared artistic expression, such as La Tène art, that linked the people of Britain and Ireland to those in modern-day France, Iberia, Germany and northern Italy. However, the concept of a Celtic people has taken a battering from archaeologists over the last thirty years, as they stress the differences between the societies that Roman writers lumped together as Celtic. Many academics avoid using the 'C' word as much as possible.

This is compounded by a dearth of evidence from Iron Age Ireland. The archaeological record in Ireland is surprisingly thin for the period between 500 BC and AD 450. Instead of revealing a flourishing civilisation, there are fewer settlements and fewer artefacts than from the periods either before or after. This was something of an Irish 'Dark Age', at least in material terms. We know that iron-making appeared in Ireland around 700 BC and La Tène objects show up around 400 years later (although only in the northern half of the country). We know the Irish of this time spoke what we now call a Celtic language, but we have no idea when this became dominant – our best guess is sometime between 1000 BC and 100 BC. We also don't know how the language took hold. There is no evidence for invasions or major population migrations during the Iron Age. Indeed, the latest genetic research indicates that the Irish population was mostly formed by the influx of people during the earlier Neolithic and Bronze Age periods. It seems that new language, technologies and art forms were imported by Irish elites who were copying symbols of prestige from Britain and the Continent, while also transforming them into something unique.

However, this Dark Age was soon followed by a Golden Age that produced an amazing library of myths, stories, poems and histories in the Irish language – the oldest vernacular literature in western Europe. Early Celtic Ireland was an oral culture, and this material was initially handed down by druids, poets and bards, who occupied specialist positions in Irish society. The arrival of Christianity brought a written culture. The old stories were first written down by Christian monks around the seventh century AD in an early form of Irish using the Latin alphabet. Many wrote on pieces of wood, which was common practice in Europe: the word 'book' comes from the German *buch*, which is itself derived from *buche*, which means beech. The earliest surviving Irish manuscripts on parchment date to the eighth or ninth century. Most of the ancient myths, sagas, poems, histories and legal texts survive in manuscripts from the twelfth to the sixteenth century, although many were copied from much older documents.

Can this early Irish literature provide a 'window on the past', and if so, how far back? The churchmen of the Middle Ages made sure to weave Christian themes into the dangerously pagan stories they inherited. To some extent, the surviving literature reflects the times in which it was written, which may have been 1,000 years after the events described. There are certainly distortions. Yet there is evidence that the epic tales of Irish literature are part of a chain of storytelling that stretches back much further, all the way to the Iron Age and pre-Christian Ireland. And literature is sometimes corroborated by facts on the ground. Navan Fort, or *Emain Macha*, in County Armagh provides one example. It features prominently in the tales of the Ulster Cycle and the exploits of Cú Chulainn as a royal site. Archaeologists have recently discovered massive ritual structures on this site dating back to the first century BC, confirming its important role at that time.

Early Irish literature doesn't provide a direct view of the past. The window is grubby and covered in the fingerprints of Christian scribes. But it does tell us something. Studying it can help us understand how pre-Christian and early Christian Irish people thought about, and used, trees.

TREES HAD AN IMPORTANT, sacred role in ancient Ireland. One legend tells how Trefuilngid Tre-eocha, a giant supernatural being with golden hair falling in curls to his thighs, came to Ireland from the setting sun in the west. Appearing before a royal assembly at Tara in County Meath, he explained the history of the island and how it had been divided into four quarters. He presented them with a branch bearing three fruits – hazelnuts, apples and acorns. When these fruits were planted, they grew into five sacred trees. Each symbolised one of the four provinces of Ireland, plus a mythical location in the centre where the provinces met. These were mighty trees, big enough to shelter a thousand warriors. At Uisneach in County Westmeath, which was the political and cosmological centre of Ireland, stood *Craeb Uisnig*, an ash that stretched 32 kilometres into the air. It was the *axis mundi*, connecting the land and its people with the heavens, around which everything turned, just as the other trees provided the axes for the provinces of Ulster, Leinster, Munster and Connacht. As this story illustrates, individual trees acted as powerful symbols for clans, chieftains and kings. The installation ceremony for a new chieftain or king was often held at a sacred tree at a royal assembly site. Sometimes the *slat na ríghe*, or rod of kingship, was cut from the sacred tree and handed to the new leader, symbolising his marriage to the land. These trees were also favoured targets for enemies. For example, a tree at Magh Adair near Quin in County Clare was a place of assembly for the Dál gCais sept, which was ruled by Brian Boru. Around 980 the then High King of Ireland invaded their territory, cut down the tree and dug it from the earth. It must have been replaced, however, because seventy years later another sacred tree on the same site was chopped down by raiders from Connacht. The fact that invading tribes made such an effort to destroy trees at inauguration sites indicates that this was a great symbolic insult to a leader and his people.

When not being fought over, trees in Celtic Ireland were a source of wisdom and poetry. Hazelnuts, in particular, had a special power. This is beautifully described in the story of Fionn mac Cumhaill. The young Fionn travelled to the River Boyne to study poetry under a man called Finnegas. The river was home to the 'salmon of knowledge', which had been fattened on hazelnuts that fell from nine magical hazel trees growing around the source of the river outside the village of Carbury in County Kildare. Finnegas had spent seven years trying to catch the salmon, as he believed it would give him great knowledge. Eventually he caught it and instructed his pupil to cook it. While cooking, Fionn burnt his finger by touching the salmon's skin and instinctively put his thumb in his mouth. He was immediately imbued with great wisdom. For the rest of his life, if Fionn was ever stumped by a problem, he only had to suck on his thumb to gain enlightenment.

As well as earning the attention of kings and poets, trees were associated with the druidic priesthood. A sacred tree was known as a *bile*. A *fidnemed* was a tree sanctuary; it denoted a clearing in a grove or wood, where ceremonial rites of a magical or religious nature were performed. The concept of sacred groves is common around the world, especially in ancient Europe. Our image of Celtic druids performing rituals in a forest is derived from the Romans. We don't know how people in pre-Christian Ireland used sacred groves. Any descriptions of rituals, including human sacrifice, were cleansed from the oral lore by later Christian scribes. But there is little doubt these sylvan sanctuaries were locations for religious rituals of some sort.

The Romans associated Celtic druids with oak trees. Pliny, the Roman historian, wrote of the Celts: 'They choose groves formed of oaks for the sake of the tree alone, and they never perform any of their rites except in the presence of a branch of it.' But this may have been a Gaulish phenomenon. There is no indication that oaks had such a central part in Irish spirituality. There were sacred oaks in Ireland but also plenty of sacred yews and ash trees too.

Perhaps the strongest evidence for the religious significance of trees in pagan Ireland was the determination of early Christian missionaries to establish churches in and around them. The Christianisation of Ireland took place in the fifth century, led by St Patrick and other missionaries. St Patrick built a number of churches beside *bile*. He founded his most important ecclesiastical centre in a *fidnemed* in Armagh. St Colmcille built his monastery in an oak wood in Derry given to him by a local king. Derry derives from the Irish *Doire*, meaning 'oak grove'. St Kevin built his first hermitage in the Holy Wood, now Hollywood in County Wicklow. He then moved eastwards into Glendalough. According to one story, he found the dense forest impenetrable, so he started to pray and the trees bent down in front of him, only to rise again after the saint passed. He lived for a number of years in a hollow tree by the Upper Lake before founding the monastery of Glendalough. (An ancient yew said to date to his time, St Kevin's Yew, stood beside the entrance until cut down by a landowner in 1835, much to the disgust of the locals.)

The founders of the early Church sought to co-opt the spiritual power of these sanctuaries, shoving the druids to the side. The transition from pagan to Christian Ireland is illustrated by the story of *Eó Rossa*, one of the five mythical trees referred to earlier. This venerable yew tree stood near Old Leighlin in County Carlow, close to the ancient seat of the kings of Leinster. A biography of St Laserian tells of how he wanted timber from *Eó Rossa* to build a church. A number of saints gathered and prayed for it to fall, but only the prayers of St Laserian managed to uproot the tree. This achieved two aims: destroying a centre of pagan ritual and providing the material for a new church. Yet by incorporating the sacred yew into the church, St Laserian was preserving its memory and redirecting, rather than confronting, the tree-worshipping proclivities of his flock.

This was a hallmark of the early Irish Church. It was distinguished by its ability to integrate pagan ideas into Christian principles and religious observances. Pagan veneration of trees and forest sanctuaries was deflected towards Christian prayer, holy trees were assimilated into the stories of the saints, and sites of pagan worship were turned into places of pilgrimage. On the Continent, the new Roman Church did its best to stamp out pagan nature worship – the more zealous bishops went around cutting down sacred trees. But the Irish Church developed into a distinct organisation, largely independent from Rome, and saw no contradiction between veneration of trees and being a devout Christian.

The story of St Laserian is also notable because he was able to pull down the venerable *Eó Rossa* without suffering any ill effects. This would have been further proof of his divine mandate, as there were strong taboos on felling or damaging sacred trees. In the twelfth century Giraldus Cambrensis, who accompanied an Anglo-Norman expedition to Ireland, wrote of the fate of archers who dared to cut sacred trees that had been 'planted by the hands of holy men' around a church at Finglas in County Dublin. 'They were forthwith smitten by God ... so that most of them miserably perished within a very few days.' Irish literature through the Middle Ages is full of stories of people who lost eyes, limbs and life because they damaged the wrong tree. Firewood that was cut from sacred trees inevitably would not burn, or the perpetrator would find his own house going up in flames. In later centuries, the stories often involved Protestant landlords or officials who cut down sacred trees after defying local customs – disfigurement or death usually followed.

These taboos ensured the survival of individual trees in Ireland over long periods, even as woodlands disappeared all around. Some echo of these taboos and venerations still exists. People still make pilgrimages to sacred trees and tie rags or leave offerings. There are hundreds of 'prayer trees', 'rag trees' or 'clootie trees' dotted around the country. They can cure all sorts of ailments and answer any prayer. For example, instead of going to the dentist, local people used to hammer coins, pins or nails into the Toothache Tree that once stood at Beragh Hill in County Tyrone. (In the long run, hammering these sorts of offerings into a tree causes metal poisoning, and many venerated trees have fallen victim to the over-zealousness of their admirers.) Most sacred trees today are found near holy wells or old monastic sites and have an explicitly Christian association. But the traditions pre-date Christianity. And there has been an uptick in veneration of these trees since the 1990s, even though the Church has lost its grip on Irish society. The person who seeks out a prayer tree in a remote corner of rural Ireland today, and ties a rag around it, is carrying on a much deeper tradition, one that dates back more than 2,000 years.

TREES WERE NOT JUST for druids and priests during the Iron Age and early medieval period. They were of great practical use as well, providing vital materials for everyday life. Metal was scarce and monopolised by the elites. Glass was unheard of. Curiously, ceramics, which had been used for the previous 4,000 years and reappeared in the Middle Ages, disappeared from Ireland and Britain during the Iron Age. This made wood even more important. We are lucky because the earliest laws of Ireland – the Brehon laws – provide lots of detail on the uses and regulation of trees.

Brehon law was a set of rules for delivering justice, originally passed down by word of mouth and later codified in text. Brehons acted as judges in public disputes and advised kings on legal decisions. (Unsurprisingly, the earliest jurists were supposed to have found wisdom by eating hazelnuts.) The law tracts were first written down in the seventh and eighth centuries AD. Most crimes could be atoned for by paying some sort of compensation.

An eighth-century legal tract, *Bretha Comaithchesa*, or Laws of Neighbourhood, regulated the use of trees and shrubs and listed the penalties for damaging or cutting trees without a landowner's permission. Trees were divided into four classes, based on practical and economic value. The most precious trees – oak, hazel, holly, yew, ash, pine and apple – formed the *Airig Fedo* or 'Nobles of the Wood'. If you cut a branch from one of these trees without having the right, you had to pay a fine of a one-year-old heifer; if you cut a fork in the trunk, you had to deliver a two-year-old heifer; and if you cut the tree at the base, the fine was a cow. The second category of tree, *Aithig Fedo* or 'Commoners of the Wood', included alder, willow, hawthorn, rowan, birch, elm and wild cherry. The legal tract specified fines of sheep, heifers and cows for unlawful cutting of these trees as well.

The remaining two categories comprised shrubs and brambles. *Fodla Fedo*, the 'Lower Divisions of the Wood', were blackthorn, elder, spindle, whitebeam, arbutus, aspen and juniper. *Losa Fedo*, the 'Bushes of the Wood', included bracken, bog myrtle, gorse, bramble, heather, broom and wild rose of gooseberry. Smaller fines were levied for damaging these lesser species.

Early Irish texts provide plenty of examples of how trees were used. One thing that stands out is the importance of hazel. As we have seen, hazel was one of the first trees to colonise Ireland after the Ice Age. Pollen analysis shows that it was more prominent in Ireland than elsewhere in western Europe for most of the last 10,000 years. Hazel grew as an understorey in oak and ash woodlands, or as pure hazel woods or scrublands on limestone areas. It was often coppiced, in which case it could survive as a shrubby tree for hundreds of years.

Despite its small size, hazel is listed in early Irish texts as a 'Noble of the Wood'. This was not only as a result of its mystical association with wisdom, but also because of the usefulness of its nuts and rods. Hazelnuts were an important element in the early Irish diet. They were highly nutritious and could be stored, making them a particularly important food during the lean, hungry winters. The strong, pliable and quick-growing rods of hazel were also vital for the construction of fences, enclosures and house walls. Walls were usually made of wattle, consisting of vertical rods interweaved with slender branches. One ingenious seventh-century house at Deer Park Farm in County Antrim had double layers of wattle, packed with insulating materials such as moss and feather, which would have made for a cosy night. One final bonus was that the nuts were used to make hazel mead, an alcoholic drink. We know that this drink was served at the banqueting hall of Tara, the mythical seat of the High Kings of Ireland, and at early Christian monasteries in Ireland.

It is unsurprising that oak qualifies as a 'Noble of the Wood'. In polls today, it is usually picked as the nation's favourite tree, and it has the moniker 'king of trees' because of its size and strong wood. Under Brehon law the oak was protected by extra regulations apart from the general fines for damaging noble trees. For example, if you cut a young oak tree, you had to pay a compensation of a two-year-old heifer.

Oak had many uses. Its bark was stripped to tan leather. Stripping bark from an oak tree that you did not own was subject to special penalties under Brehon law, as it usually led to the death of the tree. There were different fines depending on whether the bark was used to tan a man's or woman's sandals. Oak, of course, was an important construction material. The sturdiest type of fencing was described as 'oak fence', and there are many references to a type of church called a *dairthech* or 'oak-house'. The great ceremonial structure at *Emain Macha* in County Armagh was a roundhouse 40 metres in diameter, with four concentric rings of oak posts and a massive oak pillar in the centre. Through dendrochronology – reading the tree rings – we know that this central oak, which stood a whopping 13 metres tall, was cut down in 95 BC precisely. Because it burns so well, oak was also used for firewood, as charcoal for iron-making and for funerary pyres.

What is surprising now is the value placed on oak's fruit. One ninth-century legal commentary listed acorns first when describing why oaks were so precious. The acorn crop was very important to the agricultural economy because it was the staple diet of pigs in the autumn. One later text explained that a single oak could provide enough acorns to fatten one pig in a good year. Medieval texts are full of stories of beneficent kings who were blessed by abundant acorn crops (or mast years), and less fortunate royals who were punished by barren years. For example, the Annals of the Four Masters describe how the reign of the undeserving leader Cairbre Caitcheann was marked by several disasters, among them rivers without fish, cattle without milk and oak trees that produced only a single acorn. Today, Spain has its famous *jamón ibérico*, the best of which comes from black-hooved pigs grazing on acorns. Early Irish farmers used a similar approach.

The yew was not only one of the most sacred trees in Ireland, commonly found in churchyards and around monastic settlements; its timber was also highly prized for making 'noble artefacts' such as domestic vessels, spears, arrows and figurines. Indeed, one study concluded that yew was the only wood used for making stave-built vessels in early Christian Ireland. Staves were narrow strips of wood that were curved into shape and enclosed by hoops to form watertight vessels. Skilled coopers could create giant tubs for bathing or brewing, milk churns and butter churns, casks, barrels and all sizes of bowls for drinking and feasting. Although oak and ash were eminently suitable for this purpose, they were almost never used. Yew was a cultural choice, indicating its association with goodness and nobility.

A ninth-century legal commentary explains the value of the other Nobles of the Wood. Ash was used to make furniture, spear shafts, oars and yokes. It was also referred to as 'the support of a king's thigh', which probably means it was used to make royal thrones. The apple tree was, of course, prized for its fruit. Even the small sour fruit of the native wild apple would have been much appreciated by the early Irish during the late autumn and winter. Sweeter cultivated apples were probably introduced to Ireland by Christian monks, as they were well known to the Romans.

It is a little surprising to see the holly tree (*Ilex aquifolium*) listed as a 'Noble of the Wood'. Holly usually occupies the shrub layer in a forest. A staple of modern Christmases, it has prickly, bright green, leathery leaves, which it retains all year, and bright red berries. According to ninth-century Irish commentaries, holly was prized because it provided winter fodder for livestock. It was also used for the manufacture of chariot shafts.

Chariots were amongst the more impressive objects made from wood at this time. Two-wheel chariots are well known from Iron Age archaeological sites in other parts of Europe and play an important role in Irish epic literature, especially the *Táin Bó Cúailnge*, or the Cattle Raid of Cooley. They carried two people: a charioteer who drove and a warrior who fought. The term 'chariot owner' became a synonym in Irish literature for a person of high social status. This was 'the new Porsche', replacing the dugout canoe of the Bronze Age. Chariots were introduced during the Iron Age and continued in use during the medieval period.

To get around, the status-conscious chariot owner needed roads. There are lots of references in early Irish texts to different classes of roads, who was expected to build them and who was responsible for their upkeep. We even know that chariots were driven on the left-hand side. In themselves, roads could use up a lot of timber, as wooden trackways were built over wet ground. One of the longest trackways to be discovered, on the Corlea Bog in County Longford, dates from 148 BC. Extending over two kilometres, it took about 300 large oak trees and a similar number of birch trees to make. It was three metres wide, built from split planks sitting on rails, and was suitable for wheeled traffic. Its construction was a big job, involving many people. It must have left a big dent in the nearby forests.

Many of the woods would have been privately owned, especially by the chieftain, king or aristocratic class. Private woods were usually surrounded by a ditch or a wall. Celtic romanticists have claimed that woods were subject to some form of altruistic communal ownership – in contrast to the individualist property rights of later invaders from England – but this does not seem to be the case. However, all freemen in the community enjoyed limited rights in private woods, such as picking berries, collecting enough firewood to cook a meal, gathering a fistful of hazelnuts, cutting rods for carrying a dead body, or making the shaft of a spear. There were other forests earmarked for 'kindred' use, or described as 'distant' or 'waste', that were available for use by everyone or by certain classes of the tribe.

The surviving texts provide a little information on how woods were managed at this time. A seventh-century text describes how woodland could be enclosed by an 'oak fence' to keep out livestock, probably formed by partially felling young trees and bending them over to create a hedge. A law text on land values stated that the value of a wood was increased if there was access by road – something that present-day forest owners are discovering. One twelfth-century story provides evidence of the use of the 'coppice with standards' method of management, whereby a few large trees are allowed to grow to maturity while other trees are regularly cut and allowed to grow back, yielding a crop of rods every decade or so. It can be assumed that most woods were not subject to any long-term management regime. People would have gone in and cut down trees or picked fruit or nuts when they needed.

Most men, and probably some women too, were part-time lumberjacks. The large axe was a prized possession in Irish households. A ninth-century text on triads lists the three most valuable things in a house as 'oxen, men, axes', in that order. The penalty-fine for stealing or destroying somebody's axe was – surprise, surprise – a two-year-old heifer. Surviving texts provide detailed instructions on the manufacture of a proper axe. It should have triple-heated iron, hard enough not to be dented by oak or yew, the toughest woods. The axe head should be 150 mm wide with a cutting edge of 75 mm.

It seems that the felling and transport of timber could have involved a class of specialised woodmen. This was described in the seventh-century *Vita Prima Sanctae Brigitae*. In the account, a large, tall tree was felled 'by those who used to ply their trade in a timber forest' and thereafter specialised equipment or 'skilled devices' were used to drag it out of the wood to the 'appointed place'. Brehon law also specified the safety measures a tree-cutter should take before swinging an axe. He had to drive away any livestock grazing nearby, ensure there were no sleeping, deaf or witless people in the danger zone, and shout a warning – the Old Irish word for 'timber', we can presume. A tree-cutter was liable to pay a fine if anyone was injured.

EARLY IRISH LITERATURE, THROUGH its stories, poems, histories and legal tracts, gives us an insight into how wood was used, and forests regulated, from the Iron Age to the early medieval period. How did forest cover change across Ireland during this time? As we have seen, the amount of land under trees had declined to around 50 per cent at the peak of the Bronze Age. The figure may have increased slightly during a lull in human activity towards the end of the period. But the axes began to swing again at the start of the Iron Age. Through pollen analysis, supplemented by charcoal analysis and dendrochronology, we can see the patterns of woodland clearance and some of the causes. As before, this was not a linear process. There were ebbs and flows.

The many uses for wood took a toll on the forests, as people selectively cut the most favoured species. They also cut the best-quality trees, which caused genetic depletion over time by reducing the quality of regenerating trees. Yet the main driver for the clearance of woodlands continued to be farming, especially for livestock. The strong association between Ireland and domestic cattle has a very long history. At a Mesolithic settlement at Ferriter's Cove on the Dingle peninsula in County Kerry, cattle bones were found dating all the way back to 4,500 BC, well before the widespread adoption of farming in the Neolithic period. Cattle were raised on the scrubby grassland that remained after trees were felled, and herds were moved across the landscape according to the seasons. They were taken away from the home fields during spring and summer to forage on new growth in other areas, while the home fields were used to grow crops. Cattle were then brought back to graze on the grasses and herbage that grew up on the fields after a cereal harvest.

Initially, cattle were raised primarily for beef, but milking became increasingly important during the Iron Age. Caches of 'bog butter' have been found buried in wooden containers in bogs throughout the country dating from 375 BC to AD 313. This proves that people have been milking cows in Ireland for more than 2,000 years. One explanation for the absence of archaeological evidence for settlement during the Iron Age – and the lower amount of cereal pollen in the records – is that people shifted to a nomadic existence centred around cattle. Ireland was certainly famous for its cattle-rearing among the Romans. According to Pomponius Mela, there were such abundant plains of grass that cattle could eat their fill in a very short time and would burst if not restrained from devouring it.

The expansion of cattle farming led to a burst of forest-clearing in the early Iron Age, around 550 BC to 200 BC.

The iron axe was even better at cutting trees than its bronze predecessor. Pollen analysis from all parts of the country shows a decline in woodland during this period and a shift to more open, pastoral landscapes. A study of Iron Age charcoal in the midlands also detected that people were burning wood from a wider range of tree species, including light-demanding species such as birch. This indicates that people were surrounded by a more scrub-like environment, and that the main woodland canopy species had been depleted.

The decline of woodland was not solely due to human action. The bogs continued their advance as there was a turn to wetter conditions after 400 BC. Oak woodland that had survived on bogs disappeared almost completely around 200 BC. People also had to deal with the advance of the bogs. We have already seen how they built wooden trackways to cross wet ground – trackways continued to be made until the sixth century AD. Iron Age people also responded in more macabre ways. In 2003 the head, partial torso and parts of the arms of a man were found at a bog at Clonycavan in County Meath. His head had been smashed with an axe and his abdomen had been sliced open, probably in a ritual sacrifice of some type. Four other bog bodies, preserved in the cold, acidic, oxygen-depleted conditions, can be dated to the Iron Age. They appear to have been deliberately cast into bogs or lakes as part of a ritual: Irish druids did not confine themselves to forest groves, it seems. Indeed, one theory is that trackways such as the Corlea road in County Longford were built not to *cross* the bog but to gain access *to* it so that these bloody rituals could be carried out.

There was some recovery of forests during what is known as the 'Late Iron Age Lull'. In the last centuries BC and the early centuries AD cereal cultivation and pastoral activities declined, farmland was abandoned and the archaeological record thins even more. The Irish population may have fallen. While Roman civilisation was reaching its peak in other parts of Europe, Ireland entered its own Dark Age. We cannot explain this by changes in the environment, such as wetter weather or volcanic eruptions cooling the earth: these things also happened when Irish society was flourishing. It was most likely a symptom of social changes such as the fragmentation of large regional political centres.

There is an upsurge of tree pollen in the pollen records at this time. Forests made a comeback. This was secondary woodland, sometimes different from the primary forest that came before. Around Derryville bog in Tipperary, for example, all the main climax species from the old forest returned – oak, ash and elm – but a wider range of tree and shrubs grew underneath, such as holly, ivy and apple. In the north, the heavy clay soils favoured hazel, holly and birch. On the Dingle peninsula there was plenty of birch, alder, willow, hazel and holly, whereas in the west the new woodland was dominated by yew.

By the fifth century AD this 'Late Iron Age Lull' was well and truly over, and Ireland experienced another burst of forest-clearing. We are not sure why this happened, but it may have been linked to the importation of new technologies and practices from Roman Britain and the wider Roman world. Christianity wasn't the only thing Ireland adopted from the Romans. It also imported improvements in agriculture, and changes to settlements, dress, ornament, tools, weapons, literacy, vocabulary and art. Archaeologists have found evidence of Roman settlements, for example, on Lambay Island off the coast of Dublin. These were probably small trading posts, set up by the Roman merchants who plied the Irish coast. Paradoxically, we have more evidence of the settlement of Ireland by Roman British during the Iron Age than by Celtic warriors from continental Europe. Roman influence intensified with the Christianising missions of Palladius and Patrick to Ireland around AD 430–460.

This is when the history of Irish trees diverges from broader European history in interesting ways. Mid-latitude Europe (the part below Scandinavia and above the Mediterranean) experienced sustained forest loss during the Iron Age and Roman periods. In Ireland, the latter half of this period was when human impact lessened and forests recovered. Conversely, there was an increase in forest cover in mid-latitude Europe between AD 300 and 700 as the Roman empire collapsed. This was when forests started disappearing at a rapid rate in Ireland in a flurry of new settlement and building. The Irish like to pride themselves as the preservers of Roman literacy and religion, carrying the torch of civilisation while the rest of Europe descended into chaos during its 'Dark Ages'. This story can be read, to some extent, in the pollen records. If forest clearance is a mark of civilisation, the early medieval Irish bore this mark with pride.

This period also saw the appearance of one of the most iconic features of the Irish landscape – ringforts, or *ráths*. Between 30,000 and 50,000 ringforts were built between AD 550 and 850. This is more than all the settlement sites found in all of western Europe for this period. They have no exact parallel anywhere else. They were farmsteads built on mounds and surrounded by one or more circular banks and ditches, around 30 metres in diameter. They are further evidence of the deep importance of cattle to Irish society at this time. Ringforts were defensive structures built for cattle, especially dairy cows, as much as people. They were designed so that animals could be herded into the ringfort each night.

This was no bad idea, if the elaborate cattle raids described in ancient Irish legends are anything to go by. Cattle-raiding seems to have been the chief pastime of early medieval Ireland. Invading tribes may have enjoyed cutting down a rival's sacred inauguration tree, but they enjoyed stealing cattle even more. This is most famously described in the *Táin Bó Cúailnge*, or the Cattle Raid of Cooley, which is supposed to be set in the first century BC. It describes how Queen Medb of Connacht waged an epic war on Ulster in order to seize a prolific stud bull, only to be defeated by Cú Chulainn and his allies. The culture of ancient Ireland was very much a cowboy culture. This is why so many of the fines listed in the Brehon laws are paid in heifers or cows. Indeed, one story describes how an invading army had to pay a fine of 3,000 cows – unimaginable wealth at that time – for cutting down the sacred inauguration tree of the O'Neills.

Later on, forest-clearing was also driven by an expansion of arable farming. Pollen records show that cereal cultivation in Ireland increased between AD 650 and 950. This may have been linked to the introduction of a new technology, the mouldboard plough, which overturned soil more efficiently. Barley was the most important crop, with wheat, oats and rye being much less abundant. Archaeologists have discovered a number of horizontal water-powered mills used to grind cereals. One of the best examples was found at Kilbegly in County Roscommon before the building of the M6 motorway. The mill is located just 12 kilometres from the important monastic sites of Clonmacnoise. Pollen analysis showed that the mill was surrounded by a mostly open landscape. Woody vegetation consisted mainly of hazel, with small numbers of tall canopy trees, mainly oak, but also some ash, elm, alder and birch. Yew was locally extinct by around AD 870.

Around the turn of the first millennium AD, the picture of the Irish landscape that emerges from early Irish texts is one of open farmland interspersed with individual trees and small woods. Large woods were rare and confined to poor land. The author of a ninth-century series of geographical triads clearly regarded large forests as unusual in the Ireland of his day. He listed the three wildernesses of Ireland as *Fid Mór hi Cúailngi*, 'the great wood of Cooley' in County Louth; *Fid Déicsen hi Tuirtre*, 'the wood of Déicsiu in Tuirtre' probably on the slopes of Slieve Gallon in County Tyrone; and *Fid Moithrehi Connachtaib*, 'the wood of Moithre in Connacht'. In the epic stories, heroes range across the plains in their chariots. There is nothing of the dark primeval forest (*wald*) that is such a feature of German folk tales from the Middle Ages. Indeed, the various annals that record the history of Ireland boast of widespread clearing of forests for agriculture by heroic leaders 2,000 years before the arrival of St Patrick. For example, Iria the Prophet, the tenth monarch of Ireland, was supposed to have built seven palaces, triumphed in four battles and caused much of the country to be cleared of trees. There is no great longing for the primeval forest. Instead, the clearing of woodland is praised as a contribution towards progress.

How much forest and scrubland was left in Ireland by AD 1000? To answer this in any sort of quantitative way we must turn to an ambitious study by British academics that applied computer models to pollen data to estimate forest cover at a regional scale. It doesn't give a picture of the whole country, but it is the best we have. On the Dingle peninsula in County Kerry there was almost no forest left by AD 1000, just a couple of per cent of land. This is probably representative of large parts of the west that were in the grip of blanket bog. In the midlands, forest cover had declined from 35 per cent in 500 BC to just 12 per cent by AD 1000. This is probably representative of more fertile low-lying areas that were cleared for farming. Other areas saw some clearance but still retained large woodlands. For example, the area around Killarney in County Kerry went from 63 per cent forest cover in 500 BC to 38 per cent in AD 1000, while a pollen sample from Sluggan Moss in County Antrim shows that forests shrank from 65 per cent to 40 per cent of the land. Nevertheless, there were other areas that did not experience much change. Forests covered 40–50 per cent of the land around Glendalough in County Wicklow throughout this period – St Kevin's protective spells had some effect, it seems. A number of sites in County Down also show forest cover of around 40–50 per cent across the 1,500-year period.

Overall, the picture that emerges is of a treeless landscape on the blanket bogs and raised bogs in the west, the uplands and the midlands. There were scattered woodlands in the more fertile areas of the midlands, south and east. Substantial woodlands could still be found in the north of the country and in mountainous and inaccessible areas free of bog. We don't have a precise figure for the whole island, but, at a guess, it is likely that no more than 20 per cent was covered in forest or shrubland by AD 1000.

A SORT OF FAUX-CELTICISM holds that deep down the Irish are a tree-loving people, wrenched from an ancient sylvan wonderland by foreign oppressors and economic progress. This is hard to maintain. During the Iron Age and the early Middle Ages, Irish people certainly venerated individual trees and forest groves. There were strong taboos around their destruction, which ensured their survival over centuries. However, this taboo did not extend to the felling of trees in general. The Irish appreciated the usefulness of trees and harvested wood for a myriad of purposes. They were enthusiastic clearers of woodland to create land for farming. There is none of the mystical attachment to wildwoods that forms such a strong part of German culture, for example. If anything, the early Irish were a cow-loving people. You get the impression that if forced to choose between a venerable old tree and a handsome bull, they would choose the latter.

MEDIEVAL INVADERS

IN 1962 FIVE VIKING ships were excavated from the muddy bottom of Roskilde Fjord in Denmark, about 30 kilometres west of Copenhagen. One of them – *Skuldelev 2* – was a 30-metre longship. It was a war machine, a chieftain's ship capable of carrying a crew of seventy warriors. Long and narrow, with graceful curving sides, it could slice through the waves at great speeds under the power of its enormous sail or its sixty oars. *Skuldelev 2* is one of the longest Viking ships ever found, and it represents the pinnacle of Viking shipbuilding technology. And it was built in Dublin. When Danish researchers analysed the timber used in its construction, they were able to match the tree rings with the Irish oak dendrochronology record and work out that the trees had been felled near Dublin in 1042. The trees were around 250 years old, and had been carefully chosen to produce very high-quality timber.

The remains of the ship are on display in the Viking Ship Museum in Roskilde. In the mid-2000s the museum built a replica of the ship using only materials and tools that would have been available to Dublin shipwrights in the eleventh century. The project consumed a lot of wood. It required twenty oak trees with straight trunks for the keel, forestem, afterstem and planks; 250 pieces in a variety of shapes from crooked oak trees for the internal frames; three ash trees for the top planks that contained the oarports; two pine trees for the mast and yard; thirty-five pine trees for the oars; and ten willow trees to make 1,000 wooden nails. In honour of its Irish origins, the ship was named *Sea Stallion from Glendalough*. In the summer of 2007 a crew of seventy, led by blond-haired Danish skipper Carsten Hvid, sailed all the way to Dublin, where the Viking longship received an enthusiastic homecoming.

The presence of Irish oak planks at the bottom of a Danish fjord opens a new chapter in the history of Irish forestry. For the first time, in any great quantity at least, Irish timber was exported to other countries. This was the beginning of an age of commercialisation and extraction. Foreign invaders introduced new economic systems, new settlements and new trading relationships to Ireland, and integrated Ireland more closely with the rest of Europe. The Vikings came first, and then the Anglo-Normans. The existing Gaelic population assimilated these invaders to a greater or lesser extent and embraced these new patterns of trade. Irish forests continued to disappear although, as before, there were periods of regeneration when *Homo sapiens* faltered.

THE NORSEMEN FIRST APPEARED as raiders on Irish shores around AD 800. They were attracted by the lure of wealthy monastic centres such as Armagh and Clonmacnoise, which were famous throughout Europe. They started by raiding and pillaging during the summers, but they soon turned to trade and began to build permanent settlements. Some of these 'pirate bases' evolved into medieval towns. Dublin, Wexford, Waterford, Cork and Limerick were all founded by the Vikings. As time went by, the new arrivals became integrated into Irish society. Norsemen married local women and started adopting Irish names. By the time of the Battle of Clontarf in 1014, the Norse leaders were born in Ireland, they were Christian, and they were embedded in kinship networks with Irish kings. The Norse had become Hiberno-Norse.

Trade was at the heart of the Viking project. Ireland was a cashless society when they arrived. Most economic transactions in Gaelic Ireland were through gift exchange or barter, with cattle acting as the most important store of value. The Vikings introduced silver bullion (and, later, coins) to Ireland. The first coins minted in Ireland were produced by Sitriuc Silkenbeard, the Norse ruler of Dublin, in the 990s. (He died in 1042 – the same year that the oak trees were felled to make the longship *Skuldelev 2*.) The Vikings acted as economic middlemen between the native Irish and overseas markets. They acquired slaves, hides and wool from Ireland for export. Timber was also exported, either as a raw material or after being transformed into ships, vessels or artefacts. For example, some Irish timber went to the Viking colonies on Iceland and Faroes, where trees were in short supply.

Dublin, or *Dyflinn* in Old Norse, was the commercial hub of Viking Ireland. Founded during the ninth century on a promontory at the confluence of two rivers, the Liffey and the Poddle, it grew into the pre-eminent town in Ireland and one of the most important Viking towns in northern Europe. Even in the tenth century historical sources referred to it as the 'capital' of Ireland. The town grew rich from its position at the centre of a trading network that straddled the Irish Sea, then a 'Viking lake'. Its mercantile elite orchestrated the development of a substantial monetary economy that stretched out across the east of Ireland. Dublin had 3,500–4,000 inhabitants by the year 1000, and its population more than doubled during the eleventh century.

There had been small proto-towns in Ireland around monastic centres in the early medieval period, but the Vikings introduced urbanism proper to Ireland. This had a major impact on local forests. Towns were voracious consumers of wood of all sorts. We are fortunate that Dublin has been extensively excavated over the last fifty years. A trove of material has been preserved in a 3-metre layer of waterlogged deposits that lies beneath the modern city. This gives us an insight into the use of wood and the management of the city's woodland hinterland.

The Irish name for Dublin is *Baile Átha Cliath*, which means 'Town of the Ford of the Hurdles'. Before the Norse arrived there was some sort of wattle bridge over the River Liffey, part of the Slíghe Cualann, one of the five great highways that radiated from Tara. In the late tenth century the Norse of Dublin built a wooden bridge spanning the river, just east of the ford. They also built a lot of houses. There were an estimated 900 houses in Viking Dublin, each of which had to be replaced every fifteen years on average. Their walls were made of wattle, formed from inter-weaving thin branches from young hazel trees. Posts, thresholds and doorjambs were mostly made of ash or alder, again from young trees. Alder was abundant along the banks of the Liffey and the Poddle, while ash was common on the limestone-rich soils of counties Dublin and Meath. Hazel was found everywhere in Ireland at this time. More than a thousand other wood artefacts have been recovered, including bowls, barrel staves, toilet seats, furniture, ship timbers and children's spinning tops. Interestingly, most of the barrel staves found on Fishamble Street were made of yew, following the Gaelic tradition.

One of the curious features of Viking Dublin is that oak was almost never used to construct houses. This is in contrast to Waterford, where oak, often from large trees more than 200 years old, was commonly used as structural timber. One explanation is that there were no mature oak trees left around Dublin at this time, but this is contradicted by the widespread use of oak in Anglo-Norman buildings in a later period. The real reason appears to be that oak around Dublin was reserved for shipbuilding.

Dublin was a centre of shipbuilding during this period. Archaeologists have uncovered more than 400 ships' timbers from Viking and medieval Dublin, mostly drawn from nearby woodlands. From the eleventh century we have evidence that large oak woods were being reserved for royal use in other parts of the Viking world. It is likely that woodland around Dublin also came under hierarchical control, as oak was such a prized economic and strategic resource. Ships were probably built in the murky pool formed by the confluence of the Poddle and the Liffey, now buried at the back of Dublin Castle. This is where the city ultimately got its name: Dublin comes from *Dubh Linn*, which means 'Black Pool' in Irish.

Skuldelev 2 was one of the more impressive vessels built in the city at this time. It was completed around 1042 and then refurbished with new timbers from the Dublin area twenty years later. Dublin was then a maritime power: we know that the city profited from the hire of its large mercenary fleet to Anglo-Saxon kings in England. This may explain how the *Skuldelev 2* ended up in Denmark. When King Harold II was defeated by the Normans at the Battle of Hastings in 1066 his sons fled to Ireland, where the King of Leinster gave them refuge. He placed the 'fleet of Dublin' at their disposal for a planned invasion of England. The sons travelled to Denmark around this time to seek support from their cousin, the Danish king at Roskilde. It is possible they travelled on the *Skuldelev 2*.

THE VIKINGS WERE TOWNSPEOPLE. They did not venture much into rural Ireland except to raid or trade, they certainly weren't renowned farmers, and their impact on the Irish landscape was slight and mostly indirect. The same cannot be said for the next wave of invaders, the Anglo-Normans. They had Viking origins themselves, originating out of the interactions between Norse settlers and indigenous people in the Normandy region of France – a sort of Gallic version of the Hiberno-Norse. In 1066, led by William the Conqueror, the Normans invaded England and defeated the Anglo-Saxons. Over the next 100 years they consolidated their control over England and Wales before turning their sights to Ireland.

Famously, the first English invaders of Ireland came by invitation. The exiled King of Leinster, Diarmuid Mac Murchada, appealed to King Henry II of England for help to regain his kingdom. He recruited Welsh marsher barons such as Earl Richard fitz Gilbert (better known as Strongbow), who landed near Dublin in 1169. As part of their grand bargain, Strongbow married Diarmuid's daughter, Aífe, and became King of Leinster when Diarmuid died in 1171. Other Anglo-Norman warriors took the Norse towns of Wexford and Waterford. Worried that his barons would establish an independent kingdom, Henry II launched a military expedition to Ireland in 1171 to assert his authority. He received the submission of most of the native Irish rulers while there. But he was abruptly called away to deal with problems in France and, like future English kings, decided to devolve much of his authority in Ireland upon his local barons.

What did the Anglo-Normans find in Ireland? They carried out a Domesday survey of the island after their conquest, just like their forefathers in England a hundred years before, but it does not survive – the manuscript was burnt sometime before 1281. We do, however, have the writings of Giraldus Cambrensis, or Gerald of Wales. Born in 1146 of mixed Norman and Welsh descent, Gerald was related to the FitzGeralds and other Norman barons who invaded Ireland. He visited Ireland twice in the 1180s and wrote *Topographia Hibernica*, or *Topography of Ireland*, based on his travels. While full of fable and unreliable in many ways, it does contain useful information on the flora and fauna of Ireland at that time, and certainly reveals Anglo-Norman prejudices in regard to Ireland. Along with criticising their manners, dress and hairstyles, Gerald lambasted the natives for failing to cultivate more arable land, and for a lazy dependence on pastoral farming. He wrote that there were 'in some places very beautiful plains, though of limited extent in comparison with the woods'.

This image of Ireland as heavily wooded seems to have been well established in England. For example, the twelfth-century Welsh cleric Geoffrey of Monmouth, when writing about King Arthur, attributed a prophecy to Merlin in which an English king would arise to 'overthrow the walls of Ireland and turn its forests into a plain'. The Irish landscape was mostly open by this time, so this perception of a forested island is curious. The explanation may lie in the landscape the Anglo-Norman invaders were comparing it to. England by the twelfth century was one of the least forested countries in Europe, having gone through centuries of agricultural and urban development at the hands of the Romans, Anglo-Saxons and Normans. At the time of the Domesday survey in 1086, woodland and wood pasture covered just 15 per cent of the English countryside. By 1350 this figure had dropped to 10 per cent. Ireland may have been more forested in comparison, but only because England was so unusually treeless. Anglo-Norman observers such as Gerald of Wales may also have been reacting to the large areas of hazel scrub woodland then present in Ireland.

Gerald of Wales is also the first writer to complain about Irish fighters using woods for a military purpose. This would be a bitter refrain among British commanders for the next 500 years. Gerald described how the Irish would improve the defences of a wood by cutting down trees on both sides of a passage, casting some over the road, forming breast-work with others, and plashing or interlacing the lower branches of standing trees within the undergrowth to form barriers. 'In France men choose the open plains for their battles, but in Ireland and Wales rough, wooded country,' he wrote. 'There heavy armour is a mark of distinction, here it is only a burden; there victory is won by standing firm, here by mobility; there knights are taken prisoner, here they are beheaded; there they are ransomed, here they are butchered.'

The frustration of the Anglo-Normans at Irish tactics reflected a clash of military cultures. The Anglo-Normans brought with them the chivalric code of western Europe. The knightly class had developed a body of custom that regulated the conduct of war based on mutual respect, capture and ransom. The Irish fought by different rules. They carried out surprise raids and killed captives rather than ransoming them. If opposed by a superior force, they would withdraw to their 'fastnesses' – woods, bogs and mountains that were more secure. Cavalry was no use in this terrain and foot soldiers could be ambushed, especially by Irish fighters with superior local knowledge. English commentators often condemned such conduct as barbarous, but Julius Caesar had complained of the

Britons acting in a similar way when the Romans invaded their country. It is how the weak have always fought the strong.

Because woodlands were now a security threat, the new conquerors made fitful efforts to clear them. The Dublin parliament introduced an Irish Act in 1297 to address the problem. The act stated that 'the Irish' were 'trusting in the thickness of woods and the depth of the adjacent bogs' to become 'more daring in committing offences'. The royal highways were so overgrown with wood in places that it was almost impossible to pass through. Every 'lord' of the woods through which a road passed was ordered to 'cause passes to be cut and cleared', wide enough to deny cover to ambushers.

Forests played an important role during the constant skirmishes between English forces and Irish chiefs during the late medieval period. For example in 1399, when Richard II sent a large army to confront Art MacMurrough Kavanagh, King of Leinster, the Irish slunk away. One English soldier complained that the woods around Kavanagh's house were guarded by 3,000 'stout men … so nimble and swift of Foot, that like unto Stags they run over Mountains and Valleys, where we received great Annoyance and Damages'. Richard ordered the mobilisation of 2,500 local people to cut and burn the woods, but not much was achieved. Woodland fastnesses would be a feature of warfare in Ireland until the seventeenth century.

The Anglo-Normans had a greater impact on the Irish landscape through their introduction of a manorial system that was common across northern Europe. Every Anglo-Norman lord worth his coat of arms wanted to control and exploit land. Around 300 manors were established in Ireland, mostly in the east and south of the country. They were typically larger than in England: for example, manors in the Dublin region averaged 172 hectares in size, compared to 77 hectares in the London region. At the centre of each manor was the castle, built of stone as a symbol of permanence. New villages were built around each castle, attracting immigrants from England and Wales. The lords invested capital and constructed mills, fishponds, roads and bridges on their estates.

In contrast to the pastoral activities of the Gaelic Irish, the primary focus of each estate was arable farming. On average, 70 per cent of the land in each manor was devoted to growing cereals. There was a major expansion of arable acreage in the thirteenth century, much of it through forest clearance. The Anglo-Normans did not introduce arable farming and milling to Ireland as was once thought – these practices had been established for some time – but they did introduce some new technologies. And they were responsible for one deeply important innovation: 'the introduction of a commercial mind-set to agriculture'. They organised their estates with the goal of generating grain surpluses that could be traded. If cows were a store of value for the Gaelic Irish, bushels of wheat represented wealth for the Anglo-Normans.

Around the tilled fields that lay at their core, manors also had larger areas of parkland consisting of pasture, rough grazing and woodland. Medieval parks were enclosed areas of land surrounded by a wall, hedge, ditch or palings (a wooden fence) – the word 'park' means enclosure. They were used for pasturing cattle and sheep, and as pannage for browsing pigs. Instead of bringing livestock into a ringfort each night like the medieval Irish, the enclosed park protected animals from theft at all times. Many parks contained fallow deer, which were introduced to Ireland by the Anglo-Normans. They also brought rabbits to Ireland, which they bred in warrens for meat and fur. Medieval parks were usually not big enough for hunting, but were more like 'live larders', providing venison and other meat when needed.

Another important function of parks was to safeguard woodlands for the use of the lord and his tenants. The Anglo-Normans introduced woodland management techniques that were common in England at this time. Each wood was managed in two layers by dedicated foresters. 'Standard' trees were allowed to grow to maturity, when they were harvested to produce large beams and boards for construction and shipbuilding. The woodman's job was to ensure a steady progression of replacements, which could take the place of felled trees and grow into new 'standards'. At the same time, the underwood was actively managed to produce smaller timbers. This was done through coppicing (cutting a tree at the stump and allowing new rods to grow, which could be harvested every four to eight years); or through pollarding (cutting the trunk of a tree a few metres from the ground and allowing thicker branches to grow back). Again, these practices of woodland management were not new to Ireland, but the Anglo-Normans introduced more formal systems and left a documentary record. They also assigned a clear monetary value to woodlands: there are lots of examples in medieval records of underwoods being rented out for a fixed price per acre.

By producing grain surpluses and charging cash rents, the Anglo-Normans introduced a market-based and cash-based economy into rural Ireland. Lords encouraged the creation of market towns on their estates to channel marketing activity and collect more rents. They sponsored weekly markets and annual fairs. Craftspeople set up workshops in these new towns to serve the local farming community. Trade, commercial agriculture and the widespread use of coinage transformed culture and society during the thirteenth century, especially in the east and south, where Anglo-Norman rule was most established.

The Anglo-Normans also expanded the old Norse towns and built new ones. Towns such as Dublin, Drogheda, Wexford, Waterford, Limerick and Cork were adorned by the new colonists with parish churches, castles and public buildings. Dublin was by far the biggest town, with a population of 10,000 by the late thirteenth century. Kilkenny was the most important inland town, with a population of 4,500. Five other ports – Drogheda, Waterford, New Ross, Cork and Limerick – are likely to have had 2,000 inhabitants. Smaller seaports such as Dundalk, Wexford, Youghal and Galway had around 1,000 people. Altogether there were some 94,000 people living in towns at the height of Anglo-Norman rule. These towns would have needed the grain output from 228,000 hectares to supply their cereal needs. This demand stimulated the manorial estates to further expand their cultivated area.

The growing towns were major consumers of timber as well. There was much new building, and the Anglo-Normans oversaw an upgrading from wattle to wood and from wood to stone. For example, when Henry II held a Christmas court in Dublin in 1171, he was accommodated in a wooden hall constructed by his new Irish subjects on the site of the old Norse public assembly, or *Thingmoot*. This reflected an older Irish and Norse tradition of wooden halls for feasting and counsel. The English conquerors trumped this by erecting stone castles, churches and other public buildings, designed to awe. Instead of post-and-wattle houses they built sturdier, timber-framed structures. This required larger trees, so oak became the construction wood of choice. Large oak beams were also used to construct substantial riverfronts at Woodquay in Dublin, where ships could dock. Yet size wasn't everything. Medieval townspeople used humbler products from the forest. For example, species of mosses that were used for personal hygiene in latrines were specially gathered from shaded deciduous woodlands.

Just as the growing towns depended on rural estates for their grains, they drew wood from their forested hinterlands. Drogheda struggled with sourcing wood for building and fuel during this period – this may indicate that the area around the River Boyne had been largely deforested. Dublin, Waterford and other towns seem to have been well supplied. In Dublin, we know that 8-metre beams in the roof of St Patrick's Cathedral were hewn from trees from the local region. And Dublin maintained its status as a centre for shipbuilding: King Henry III ordered the city of Dublin to build a 'great galley' for the defence of Ireland. After it had been delivered in 1233 he ordered another one in 1241.

We have records of a large-scale rod-cutting project around Dublin in 1303. In that year William de Moenes was tasked with providing rods to make hurdles, which would assist the Earl of Ulster in crossing rivers during a planned expedition to Scotland. De Moenes oversaw cutting at seven different locations north and south of the Liffey. Altogether, the men cut between 5,000 and 10,000 rods over the course of a few days, an impressive feat of organised woodland management. The rods were brought by cart, packhorse and boat to the priory of All Saints, the future site of Trinity College Dublin.

As well as increasing the flow of timber around Ireland, the Anglo-Norman conquest was followed by greater export of timber *from* Ireland. This started early. Circa 1175, Irish oak boards were used in the rebuilding of Canterbury Cathedral in England after a fire. The boards came from trees that had been felled in the Dublin area around 1170, only a year after Strongbow first landed in Ireland. The authorities at Canterbury Cathedral must have been happy customers, because they bought another 434 Irish boards in 1235 and more 'bordi de Hybernia' twenty years later.

Irish timber played an even more important role in the construction of Salisbury's new cathedral. Timber equivalent to forty oak trees was ordered from Ireland. There is a surviving English royal patent roll entry for 15 June 1224 guaranteeing 'safe conduct along the coasts of England and Ireland for the ship in which William of Dublin conveys timber for the fabric of the church of Salisbury'. Recent dendrochronological research reveals that the Irish timber not only arrived but still survives in the roof of the cathedral. It was derived from large oak trees, about 300 years old, felled in the spring of 1222 in the Dublin area. We also know that Irish oak was used in the construction of royal castles at Marlborough, Winchester and Haverford. This shows that there was an organised forest industry in Ireland around the middle of the thirteenth century, with a good reputation in England of being capable of fulfilling significant orders for timber.

This trade in timber was just one part of a flourishing economy under Anglo-Norman rule. Ireland not only became a grain basket for the English kingdom; there were also substantial wool exports to Flanders, where it was used to make cloth. Coins circulated widely in Ireland in the thirteenth century as the Anglo-Normans developed a market-based, capitalist economy. Italian merchant bankers arrived in Dublin to supervise the mint, engage in import-export trades and lend money to government, clergy, aristocracy and townsmen. The Irish population grew, partly through immigration and partly through an improvement in living conditions. Estimates for the size of the Irish population around 1300 range from 700,000 to 1.3 million, compared to a population of around half a million in 1000.

What effect did the Anglo-Norman conquest have on forest cover in Ireland? One reasonable estimate is that roughly 20 per cent of Ireland was covered in woodland when Strongbow and his friends arrived in the 1170s. This was somewhat higher than in England. Pollen analysis shows that forest cover declined over the next 150 years, accompanied by a distinct increase in cereal pollen. There is no way of assessing the total amount of woodland in Ireland during this period. All we have are localised pollen studies that tell the story of specific areas. Nonetheless, at the peak of Anglo-Norman influence at the end of the thirteenth century, it is likely that forest cover in Ireland had decreased, perhaps to around 15 per cent.

THE HIGH POINT OF Anglo-Norman rule and the manorial system was followed by a steady decline. Ireland was hit by a series of calamities in the early 1300s. A plague afflicted cattle in 1315–16, wiping out a large part of the Irish herd. A series of harvest failures between 1314 and 1322 led to a general famine in Ireland and northern Europe. Climate change had some role in this: Europe had benefited from an unusually warm period starting at the end of the first millennium, which helped the

Vikings reach Greenland and beyond, and supported cereal farming on the cool Atlantic fringes. This was followed by the Little Ice Age between 1350 and 1850, a period of cold weather caused by oscillations in solar activity that threatened crops in marginal areas and caused the River Thames to freeze. With unfortunate timing Edward the Bruce, brother to the King of Scotland, invaded Ireland in 1315 and rival armies marauded around the country for three years before his final defeat. Yet all this was nothing compared to the appearance of the bubonic plague in Ireland in 1348–9 when between a third and half of the Irish population succumbed to the Black Death, wiping out much of the population growth of the late medieval period. Manors, villages and settlements were deserted.

The effects of this collapse can be read in the pollen record. Cereal farming ceased in many parts of Ireland, sometimes for the next 150 years. As the population collapsed, people abandoned more marginal areas and concentrated on better land. This allowed woodland to regenerate. For example, one study shows that there was an expansion of hazel scrub in the midlands in the late medieval period, although oak, elm and ash never fully recovered from the earlier clearances. We can also read this woodland revival in oak tree-ring records. Few timbers used in buildings are dated from between 1306 and 1350, indicating a pause in new construction at this time. Conversely, many timbers used in buildings in later centuries are from trees that started growing around 1350. They came from the seedlings that had a chance to grow on abandoned pasture or arable land, indicating a resurgence of oak woodland at this time. The rampage of the Black Death and the other Horsemen of the Apocalypse was a disaster for *Homo sapiens*, but it provided a welcome respite for Irish woodlands. Woodland cover, if we include hazel scrub, may have increased to the level witnessed by Gerald of Wales when the Anglo-Normans first arrived.

This was also a period of political change. English power gradually leaked away during the fourteenth and fifteenth centuries. Gaelic chieftains raised armies and regained control of large parts of the country. By 1470 English royal authority and English common law, which had once embraced two-thirds of the island, was only felt in the Pale around Dublin, in parts of Munster and in isolated pockets around walled towns in the west and north-east. Beyond lay what the English called the 'wild Irish'. There were constant skirmishes between English and Gaelic forces across an extended frontier zone. The Anglo-Norman lords had also become increasingly Gaelicised. They spoke Gaelic, intermarried with Gaelic noble families and occasionally joined in rebellions against the English Crown.

The decline of English rule was accompanied by a Gaelic revival – politically, culturally and economically. Gaelic Ireland produced very few manuscripts between 1200 and 1350, but there was a flourishing of writing after that, which points to the increased security enjoyed by the learned hereditary families. Economic activity spilled out of the manors to the broader population. There was a shift away from arable farming to livestock, as demand for meat grew and prices for wool and cattle increased in England. One new feature was the *creaght*, a mixed herd of cattle and sheep that was moved through areas of woodland and rough grazing by itinerant herders who never stayed in one place for long. Gaelic merchants embraced the commercial systems that had been introduced by the Anglo-Normans and increasingly carried out trade in Irish towns and ports. Around 3,000 tower houses were built by Gaelic and Anglo-Irish lords during this period, a sign of an emerging upper-middle class who were able to generate sufficient economic surplus to fund an expensive construction spree. The country's population recovered from its fourteenth-century collapse, and had reached around 1 million by 1500.

There was also a boom in foreign trade in the 1400s. This time Gaelic merchants played as important a role as the old Anglo-Norman families, although by then it was often hard to tell one group from the other. This trade included Irish timber. Scottish lairds sourced timber from Wexford and Wicklow, usually via Dublin, to construct galleys. The most important part of the timber trade involved oak staves, which were in high demand in Europe for the manufacture of barrels. They went to north-west Spain, the Canary Islands, Madeira, Portugal and France and came back as barrels filled with wine. Most staves were shipped from Waterford and Limerick, because these towns had good harbours and nearby woods. For example, large quantities of oak were floated down the rivers from the Shillelagh woods on the Wexford–Wicklow border. (Interestingly, this medieval trade in wood and alcohol is the reverse of what happens today: Irish whiskey-makers now source staves from Spain to create wooden barrels because there is not enough oak available in Ireland; they then ship bottles of whiskey back to eager customers in Spain.)

There was an 'air of fraud, subterfuge and, indeed, treason' associated with this Irish commerce because England and Spain were often at war. The Crown struggled to control the ports and to collect duties. It was a sign of the weakening of English rule in Ireland and the deepening connection between Ireland and Spain at that time. Spanish ships were a frequent presence at Galway, memorialised through its famous Spanish Arch. Christopher Columbus visited Galway in 1477 during his early career as a merchant seaman, and a man from Galway sailed with him on his first voyage to the Americas in 1492. One English cartographer, perhaps reflecting security concerns felt in London, identified the Atlantic waters off south Munster as 'the Spanish Sea'.

THE EMERGENCE OF A flourishing Irish trade with Spain shows how Ireland was becoming increasingly interconnected with a broader European economy. This process had been started by the Vikings, deepened by the Anglo-Normans and carried on by Gaelic and hybridised Anglo-Irish merchants. This had an impact on Irish woodlands. Irish oak became a prized commodity, whether transformed into ships, exported as boards for English cathedrals or sent to Spain as staves for wine barrels. It stimulated more active woodland management, through coppicing, but also woodland clearing. During this 500-year period the amount of Irish woodland fluctuated. After limited impacts during the Viking era, apart from the immediate surroundings of Dublin, there was substantial forest-clearing during the late twelfth and thirteenth centuries. Woodland regenerated during the second half of the thirteenth century after a series of calamities led to a population collapse, but there was further clearance during the 1400s, as the population rebounded. The proportion of the Irish landscape covered by woodland was probably around 15 per cent by 1500. This cycle of deforestation and regeneration was well known to Gaelic tradition, being encapsulated in a saying: *Teóra h-uaire do cuir Eire, Teóra monga, agus teóra maola dhí* ('Three times Ireland was cultivated: thrice wooded, and thrice bare').

The closer links between Ireland and Spain were also an important trigger for the final English conquest of Ireland. A resurgent English Crown could not tolerate the growing Spanish influence in its backyard. Henry VIII initiated a Tudor conquest of Ireland that would grind forward over the next 150 years until it was absolutely complete. The Gaelic revival was short-lived. This is the most contentious era in the history of Irish forestry. It pitches nationalists against revisionists in an argument about who should be blamed for the disappearance of Irish woodlands. In the next chapter, we will try to hack through the thickets, climb up on the hills and get a good view of the Irish landscape before and after the English (re)conquest so we can see how woodlands changed.

CONQUEST AND COMMERCE

WOLFWALKERS IS A BEAUTIFUL film from the Irish animation studio Cartoon Saloon. Set in the 1650s after the English reconquest of Ireland, it tells the story of a girl in Kilkenny whose father is a wolf-hunter tasked with exterminating wolves from a nearby wood. In the animated landscape the sharp, angular, grey lines of the town and its surrounding fields, and the jagged tree stumps poking out of newly cleared land, contrast with the flowing, swirling, verdant shapes of a native woodland that seems to go on forever. This sylvan paradise is home to a pack of wolves and a family of 'wolfwalkers', humans who take on the form of wolves at night. The Lord Protector – a figure based on Oliver Cromwell – is intent on destroying the woods. He tells the townspeople that they 'must conquer the wild ... what cannot be tamed must be destroyed'. In the end, the wolfwalkers ambush the English troops among the trees and escape to a more remote, richly forested part of the island. *The Guardian* described it as an 'eco parable' in which the wolves represent a mystical spirit of Irishness, menaced by a fanatical English oppressor.

This animation reveals two widely held beliefs about the Irish landscape. The first is that Ireland was cloaked in woodlands before the start of the Tudor reconquest in the sixteenth century. The second is that the deforestation of Ireland was the fault of English invaders, directed by Queen Elizabeth I, Oliver Cromwell and their henchmen. A common version of this story is that the trees were cut down to build the Royal Navy. It is an opinion you are likely to hear from the man or woman in the pub, the driver of a Dublin taxi or most people who have been through the Irish school system. This was long the view of Irish nationalist historians as well. They contrasted a sylvan Gaelic world, where a squirrel could travel from coast to coast without ever touching the ground, with the bare, devastated landscape left by English soldiers and settlers. For example, writing in the 1930s, in books like *No Forests, No Nationhood* and *The Rape of Ireland*, John Mackay described how Ireland's forests were eradicated alongside its liberties in a 'holocaust' of destruction. According to this view, we should look to the English reconquest of the sixteenth and seventeenth centuries to understand why Ireland became a country with no trees.

This was certainly an era of great change in Ireland. The Gaelic revival was brought to an end by the more muscular policy adopted by Henry VIII after 1534. The Crown raised armies, extended English sovereignty throughout the island and established new 'plantations' with immigrants from Britain. The Reformation introduced a religious divide that gradually pushed the descendants of the Anglo-Norman settlers – now known as the 'Old English' – into alliance with the Gaelic lords. There was a series of rebellions and wars, sometimes involving continental powers such as Spain, sometimes fuelled by instability in England, as parliament fought king and rival kings fought each other. Each outbreak resulted in ever-worse defeat for Gaelic and Catholic interests, which led to the confiscation and redistribution of land to 'New English' (and Scottish) colonisers on a massive scale. This was not only an epochal change in an Irish context; it was striking even in European terms. One scholar concludes that Ireland experienced 'the most rapid transformation' in economy, society and culture of any European country in the seventeenth century.

In theory we should have a much clearer picture of the evolution of Irish woodlands because there are more extant written records for this period than for any time before. English army commanders left reports on their campaigns against Irish insurgents, English planters wrote letters on their attempts to establish themselves, government officials prepared schemes for the administration of newly controlled territory. Land surveys were carried out as part of the confiscations and settlements, along with maps showing existing woodland.

Nonetheless, the picture is fuzzy and conclusions are contentious. In a recent book, Nigel Everett writes that 'the history of Irish woods is replete with uncertainties, misapprehensions and elements of paradox'. Depictions of Irish woodlands in English records can be contradictory. There are sometimes biases at work, as military commanders try to justify their limited progress or settlers manipulate perceived land values to get a better deal for themselves. The record is patchy, especially because many documents were destroyed when the Irish Public Records Office, attached to the Four Courts in Dublin, went up in flames at the start of the Irish Civil War in 1922. And descriptions of woodlands by contemporary observers do not always match scientific data from pollen records.

Over the last twenty-five years the accepted nationalist wisdom has been challenged by people who say that the forests were never as extensive, nor the destruction ever as great, during this period. This alternative view reflects bigger debates in Irish history as 'revisionists' replace traditional narratives of the clash between England and Ireland with more subtle interpretations. According to Professor Valerie Hall of Queen's University Belfast, 'the fate of the Irish woodlands at the end of the medieval period remains one of the most contentious in the history of the recent Irish landscape' – which is fighting talk in the world of historical ecology. Where does the truth lie?

If we want to unravel this knotty mystery, the starting point has to be an assessment of the extent of forest cover at the start of the English reconquest. In her book *Irish Woods Since Tudor Times* Eileen McCracken estimates that around one-eighth, or 12.5 per cent, of the country was covered in woodland by 1600. It is fifty years since her book was published and there have been many alternative opinions expressed since then, but her original figure is probably about right, with one proviso. It is almost the same forest cover as Ireland has today, which means not a lot – Ireland has one of the lowest amounts of forest in Europe. It also seems right when compared with the situation in England. By the end of the Middle Ages, around 10 per cent of England was covered in woodland; based on assessments by contemporary English visitors, it is hard to imagine that Ireland had less woodland. Yet this was not all tall, dense forest dominated by mature oak or ash; there were large areas of scrub woodland too. Different definitions of what constitutes a woodland may explain some of the discordant opinions on what existed at the time.

The surviving woodlands were not evenly distributed. There were few large woods left in the well-developed Pale, which included the counties of Dublin, Kildare, Meath and Louth. The western seaboard was mostly open because of extensive bogs, as was the midlands, although there were still notable woodlands in places like Offaly. The largest and densest areas of woodland lay around Lough Neagh, in the Erne basin, along the Shannon, in the river valleys and mountains of Munster, in Leitrim and on the eastern slopes of the Wicklow and Wexford hills. Some of these woods were vast by present-day comparison, covering thousands of hectares each: a map from 1580, for example, indicates that Glenglas forest on the Limerick–Cork border was 19 kilometres long and 12 kilometres wide. There were smaller pockets of woodland dotted all around the country. Kenneth Nicholls, an expert in early modern Gaelic Ireland at University College Cork, pointed out that forests were often found on the boundaries of baronies and lordships, especially between Gaelic lordships and areas of Anglo-Norman control – there was an element of the frontier about them.

Every woodland has its own story to tell. We can get an insight into the changes wrought on the Irish landscape by looking at the fate of two woodlands at each end of the island: first, the forests around Lough Neagh in County Derry that were, briefly, home to a powerful Gaelic king who had the best chance of reversing the tide of English invasion; second, the woodlands of the river valleys of Cork, Kerry and Waterford that were exploited by a series of English adventurers and planters from the late 1500s. By peering into these woods, we can glimpse some of the trauma and transformations that affected the Irish landscape in this violent period.

ULSTER WAS THE MOST heavily forested, and most Gaelic, region of Ireland in the late sixteenth century. Apart from a few settlements on the east coast, it had largely escaped Anglo-Norman rule in the medieval period. And its Gaelic lords grew in boldness as English authority waned. The most powerful of these lords, the man who came closest to reversing the tide of Tudor conquest, was Hugh O'Neill. He had the sort of allegiance-shifting career that was typical of his time. Born in Dungannon around 1550, he was reared near Dublin and served in the English army in Ireland. After becoming Earl of Tyrone with English support, he increasingly baulked at the encroachment of Crown authority on his lands. In 1595 he was inaugurated The O'Neill in the style of the old Gaelic kings and went into rebellion. Aided by his son-in-law Red Hugh O'Donnell he raised an army, confronted the Crown forces and carried rebellion through the country, even to the fringes of the Pale.

In the Nine Years' War that followed, woodlands played an important military role. Thomas Burgh, the English lord deputy, complained in 1597 that O'Neill had 'fortified all the bogs' and 'barricaded the passes of the woods'; the Irish chief would emerge from his lair for brief skirmishes, and then retreat back to his fastness. As before, Irish forces set up ambushes in forests, digging trenches across roads and plashing the wood on both sides for cover. This nullified the great advantage of the English army, their cavalry, and resulted in some successes. For example, in one ambush in a wood near Portlaoise, the hapless Earl of Essex lost 500 men in what became known as the Pass of Plumes because of the abundant loss of English helmets. The Calendar of State Papers for 1601 recorded how 'the woods and bogs are a great hindrance to us and a help to the rebel, who can, with a few men, kill many of ours'. Eventually, however, the Crown forces, led by Essex's successor, Lord Mountjoy, adopted Irish tactics. They learned to dig trenches, plash woods themselves and move lightly across terrain, carrying out their own ambushes.

The high point of the war occurred in 1601 when Spain landed an expeditionary force in County Cork. O'Neill marched south to join them but was unexpectedly defeated at the Battle of Kinsale. He slunk back to Ulster, burnt his capital at Dungannon and took refuge in the ultimate fastness, the great woods of Glenkonkyne and Killetra north-west of Lough Neagh in County Derry. These were one of the largest areas of woodland in Ireland. When the English army arrived in pursuit, one of its commanders, after climbing a hill to survey the wilderness, recorded that 'they were 10 mile broad, and 20 mile long, all covered with thick wood' – an area of more than 50,000 hectares. Rather than venturing inside they surrounded the forest and tried to starve the Irish fighters out. O'Neill avoided capture and, amazingly, was able to negotiate a surrender in 1603 that led to a royal pardon. But his position was eroded by Dublin Castle's legal and administrative advances and he left Ireland for good in 1607, along with other Gaelic leaders, in what became known as the Flight of the Earls.

The departure of the Gaelic lords was declared treasonous, leading to the confiscation of their lands and the 'plantation' of Ulster. The English government encouraged the City of London Companies to invest in this project and put forward the untouched woodlands as one of the major attractions. The promoters referred to 'the goodliest and largest timber woods' around Glenkonkeyne and Killetra, suggesting the timber could be brought to the sea by the River Bann and either exported for easy profit or used in construction in the province. This timber would be needed for the development of a new city of 'Londonderry', a name that planted an English stamp on the

Gaelic word for 'oak grove', *doire*, echoing the landscape that had been so precious to St Colmcille 1,000 years before.

The new settlers from England and Scotland regarded clearing trees as a patriotic duty. They sought to introduce 'civility' to the barbarous Irish by reshaping the landscape, enclosing fields and increasing the area under tillage or manured pasture. There was also a security imperative. As the Protestant planters settled on the best land, the Gaelic Irish were forced onto hills and more marginal land. Some kept their weapons, took refuge in the woods and launched raids on the new towns and farms. The English called them 'woodkerne'. The term referred to a lightly armed Gaelic soldier or bandit, usually with long hair and a shaggy woollen cloak, who emerged from the woods to terrorise the settlers. This figure became a semi-mythic bogeyman in the English imagination. He appears more than once in Shakespeare: in *Henry IV*, written around 1597, the Duke of York is warned of 'the uncivil kerns of Ireland' who 'temper clay with blood of Englishmen'. In 1610 one English planter identified the wolf and the woodkerne, hiding in the 'inaccessible woods', as the two greatest threats to the new settlers in Ulster. Hunting wolves and woodkernes from the woods became a kind of sport to English planters.

The planters also exploited the woods of Glenkonkeyne and Killetra for profit. The Crown granted the City of Londonderry control of the woods in perpetuity under the condition that any timber would be used to develop the Ulster plantation and not sold as merchandise. Thousands of oak trees were cut down to construct new buildings in the city and houses on new estates. According to one source 150,000 oaks, 100,000 ash and 10,000 elm were used in the building of Londonderry. But many trees were also felled to produce staves for export to Britain and continental Europe, where they were used to make barrels and other types of wooden casks. One entrepreneur was said to have cut between 50,000 and 70,000 pipe staves per annum for seventeen years. There were 100 timber-workers on the River

Bann in 1611, mostly English or Scottish immigrants. They included thirty-two fellers, twenty lath tenders, nine sawyers, eight wainsmen, four timber-squarers, four shipwrights and three overseers. There were fifteen men rafting timber down the Bann, nine men working the cots and fifty men carting timber to the river.

The enthusiastic felling along the River Bann earned the attention of the Privy Council and King James himself. The 'great woods' of Ulster had been preserved for Crown use, with an eye to supplying the strategic needs of England. In 1612 King James reprimanded the Londoners for having 'committed great havoc in the woods of Glenkonkyne and Killetra', cutting staves and 'exporting them to foreign parts' – including to the arch-enemy Spain – 'contrary to the laws'. This was a frequent refrain over the next twenty years. In 1635 the Londoners were even prosecuted before the Star Chamber at Westminster for their mishandling of the Ulster plantation, with their profit-hungry clearance of woods one of the principal charges. Yet these attempts at royal control never had much effect on the ground. By 1650, after a half century of uncontrolled felling, the Lower Bann was so stripped of trees that it was a mostly patchy scrub landscape choked with tree stumps. These were later grubbed out to create the open fields of today. Only one pocket of native woodland survives, the 26-hectare Portglenone Forest now managed by the Northern Ireland Forest Service.

So far, this story seems to confirm the narrative of an arboreal holocaust brought about by English conquerors. But recent pollen analysis provides a twist in the tale. On the banks of the Lower Bann valley, well within the bounds of the great forest of Glenkonkeyne, there are a number of lowland bogs whose top metre of peat contains pollen grains from the past 1,000 years. Study of this pollen reveals that oak woods were never as dense in this region as historic documents imply. Moreover, there is no tell-tale drop in oak-pollen values in the seventeenth century that would indicate widespread felling at this time. Instead, the pollen record shows a steady decline in woodland that stretches back over 1,000 years and continued after the seventeenth century. There is no doubt that the woods were exploited at this time, but it was an intensification of a long-running trend rather than something new and foreign. The Londoners who invested in the plantation soon came to this conclusion: they complained that Glenkonkeyne and Killetra, far from being virgin forests, had been 'so spoiled by the people of the country in late years, that the best part was cut down and purloined away'.

There *is* a clear signal in the pollen records for one major change in woodland in the seventeenth century – the rapid and extensive clearance of hazel scrub. This low hazel woodland had been a distinct feature of the Irish landscape and an important part of Gaelic life for hundreds of years. It provided rough grazing for roaming herds of cattle and sheep; its rods were used in housebuilding; hazelnuts were not only an important source of nutrition in winter but had long been associated with wisdom and poetry in Gaelic myth. The presence of large areas of scrub woodland was one way in which the landscape of Ireland differed from England, where there were clearer boundaries between managed woodland on the one side and open fields for cultivation or pasture on the other. It also helps explain why it is so hard to judge the size of Irish woodlands from contemporary records. When drawing up their maps, surveyors concentrated on timber woods, as they were regarded as 'profitable'. The larger areas of scrub were often ignored or bundled together with bogs and 'wastes' within an amorphous category of 'unprofitable' land. McCracken's estimate that woodland covered one-eighth of the island around 1600 is probably only correct if we include this degraded, scrubby woodland in the definition.

The planters sought to eradicate this landscape by creating tidy, enclosed fields with smooth grass and clear hedgerows. This went hand in hand with an intensification of land use and the commercialisation of agriculture. In some ways the clearing of hazel scrub symbolised the end of the old Gaelic order. It was the end of a tradition of silvo-pastoralism (meaning the grazing of livestock within woodlands) that stretched back to the Ireland of ancient legends. The world of Cú Chulainn had disappeared.

HUGH O'NEILL HAD EVADED capture in the fastness of Glenkonkeyne. The leader of an earlier rebellion in the south of Ireland was not so lucky. Munster was in an almost constant state of warfare between 1569 and 1583 as the earls of Desmond, one of the great Anglo-Norman families that had become Gaelicised, rebelled against Tudor centralisation. After the rebels were defeated in the field they retreated to the forests. In 1582 the commander of the Crown forces in Munster advised that they should garrison 2,000 footmen and 300 horsemen in the woods in the summer so that the rebels could be 'beaten out' and killed in the open. Gerald Fitzgerald, 14th Earl of Desmond, was pursued for more than two years in the wilds of County Kerry. He was eventually hunted down, killed and beheaded by the O'Moriarty family in Glanageenty wood near Tralee. (This wood was described by an observer as six kilometres long and three kilometres wide – an area of more than 2,000 hectares.) His head was sent to Queen Elizabeth, who, according to a courtier, stared at it for hours before having it mounted on a pole on London Bridge.

The Desmond Rebellion was followed by massive land confiscations and the plantation of Munster by English settlers. Around 3,500 English soldiers, officials, merchants and country gentlemen were given almost 120,000 hectares of land. One of the recipients was Walter Raleigh. He was the quintessential Elizabethan adventurer: a poet, courtier and soldier who would go on to explore the Americas and popularise tobacco. In 1586, after playing an active role in the Irish wars, he was granted 16,000 hectares of ex-Desmond lands around Lismore in County Waterford and Mogeely in County Cork.

Raleigh immediately set about exploiting the woodlands. He established a company with some local business partners to export timber and staves to England, Madeira and the Canaries.

He later wrote that his company had extracted 700 tonnes of timber and exported 340,000 pipe staves over three years, generating £1,000 in customs for the Crown. They had left the 'great timber' standing, because it could only be transported at great charge. But it was an unsatisfactory business: he claimed that his company had invested £5,000 and 'not returned the half'. He ended up in litigation with his local partners and sold all his Irish lands in 1604 for the modest sum of £1,500.

That the Munster plantation was a capitalist and colonial enterprise is illustrated by the involvement of the East India Company. After receiving a royal charter in 1600 to exploit trade with Asia, the company established itself in County Cork ten years later. The officers brought in 300 English settlers and built an ironworks, a timber-processing facility and a shipbuilding site at Dundaniel, at the confluence of the Bandon and the Brinny rivers. They purchased or leased thousands of hectares of woodland to supply timber. Like Raleigh, they manufactured staves and sent timber to England. Continuing a piratical tradition that stretched back to the Vikings, they also built ships. Two large ships of around 500 tonnes each were launched at Dundaniel in 1613, and ships built of Irish timber made voyages across the Indian Ocean. When the British House of Commons considered a petition from the East India Company in 1639 listing its many services to British prosperity they included 'the tonnage employed by the company to fetch timber, and pipe staves out of Ireland'.

There was much trading of land between the early planters as well as attempts to acquire or seize additional land from Gaelic chiefs. Both Raleigh's estate and the land on which the East India Company's settlement was built ended up in the hands of the same man – Richard Boyle. Boyle has been described as a 'model planter', the 'poster boy' for the English colonial enterprise in Ireland and 'the most efficient and single-minded English pursuer of Irish land in his lifetime'. He had arrived in Ireland from Kent in 1588 with little to his name – just £27 in his pocket (around £8,000 in today's money), some fine clothing and his 'rapier and dagger'. Through positions in the administration and two judicious marriages, he began to accumulate land. His big break came in 1604 when he acquired Raleigh's 16,000-hectare estate. By legally challenging insecure titles, buying out struggling Gaelic Irish and Old English landowners and lending against and then foreclosing on properties, Boyle amassed a huge estate across the country. By 1641 his rental income was £18,250 per annum (more than £2 million in today's money) and he was reputed to be the richest man in the British Isles. He was created Earl of Cork in 1620. Craving the acceptance of his new peers and attempting to silence the whispers about his lowly social origins, he spent a fortune on the trappings of nobility and developed a magnificent residence at Lismore Castle in County Waterford. A surviving oil painting shows him with balding pate, arched eyebrows and a goatee beard, surrounded by a broad fan of ruffles, looking a little like Ming the Merciless in the comic strip *Flash Gordon*.

According to a recent study, Boyle is 'possibly best understood as an entrepreneur or tycoon in a colonial setting'. The forests of the Blackwater, the Bride and the Bandon river valleys were, to him, an important economic resource. He exploited woodland on his own estates while also purchasing timber from surrounding landowners. He continued the long-standing manufacture and export of staves from woodlands across Munster. His records show transactions involving 4 million staves (around 14,000 cubic metres of wood) between 1616 and 1628. He promoted harvesting of bark for tanning, which was an important industry at this time. Boyle was a proto-industrialist who developed mines and built factories. He established a glassworks at Ballynageral in County Waterford where some of the best-paid craftsmen in the country produced eighteen cases of glassware a week. This required charcoal for fuel, while the alkali used in processing glass came from ash trees.

By far the biggest consumers of wood were the ironworks

that Boyle established in multiple locations. Iron, although low quality, is widespread in Ireland, along with limestone, which is used as a flux in iron-making. Charcoal was essential to the chemistry of smelting: it took 2.25 tonnes of charcoal to make 1 tonne of bar iron. Because of the costs of transporting timber, ironworks were usually situated close to woods. One hundred and sixty ironworks were established in Ireland in the seventeenth and eighteenth centuries. These were big industrial developments, requiring thousands of pounds in investment and hundreds of workers, mostly imported from England or the Continent. The wood to make charcoal was much cheaper in Ireland than in England, hence the attraction for English ironmakers and investors like Boyle.

Boyle oversaw a transformation of the landscape of Munster during this period. Like the planters in Ulster, he was obsessed with the notion of imposing 'civility' on the land. He established orderly fields and boundaries, in essence taming the countryside and facilitating adoption of British farming practices. He also built roads, bridges and towns. Bandon in County Cork was his showpiece, but many other towns carry the same template of Protestant church, town square, Big House and shop buildings. He created a colonial militia – a private army in effect – composed of his tenants and followers, and used this to quell dissent and to impose English law. The common-law framework of property rights underpinned this new society. Looking back in 1632, the Earl of Cork proudly wrote: 'The place where Bandon Bridge is situated is upon a great district of the country and was within the last twenty-four years a mere waste bog and wood serving as a retreat and harbour to woodkernes, rebels, thieves and wolves and yet now (God be praised) as civil a plantation as most in England.'

These words were premature. Boyle's world would soon come crashing down. In 1641 the simmering resentment of the dispossessed sparked a rebellion that began in Ulster and quickly spread across the country. Boyle's lands in Munster were overrun. His second son, Lewis, was killed by rebels at Liscarroll in north County Cork in 1642 and his extensive family – he had fifteen children – fled to England. The Earl of Cork died in 1643 with his Irish empire in ruins.

There followed what has been called 'perhaps the bloodiest and most tragic period in Irish history'. A Catholic Confederacy of Gaelic Irish and Old English took advantage of the vacuum created by civil war in England and tried to exert control over Ireland. But after years of skirmishes, they were eventually crushed by Cromwell's savage campaign in 1649–50. The English parliament raised money for yet another reconquest of Ireland through an Adventurers Act that offered up 2.5 million acres (1 million hectares) of land, to be confiscated from Catholic rebels in exchange for loans. When victory was secured, Cromwell ordered the largest seizure and redistribution of land in Irish history. The cycle of plantation, woodland exploitation and agricultural development started again. Richard Boyle's sons returned, reclaimed their estates and added to them. The Boyle dynasty was secured.

THE FATE OF THE WOODS in Ulster and Munster points to a few conclusions. First, the woods of Ireland were not felled to build the English navy. Nor were they cut down on the instructions of Queen Elizabeth or any other English ruler. Few Irish timbers ended up in England's warships. Most English ships were built in English shipyards using English oak. It is fitting that the largest ships built in Ireland at this time were built by a private company, the East India Company. This illustrates how the exploitation of the woodlands in Ireland was driven by adventurers and entrepreneurs operating independently of official policy and in pursuit of private profit.

Indeed, the Crown made repeated attempts to regulate the felling of Irish woodlands, so as to reserve large timbers for future use. In 1583 Queen Elizabeth ordered that Crown woods in Ireland growing 'commodiously near any river having portable recourse to the seas' must be 'carefully reserved and preserved' for the 'maintenance of our navy' and 'our other buildings'. An appendix to the official *Survey of the plantation of Munster* complained that some of the planters had 'committed great waste of woods and timber by iron works, making of pipe staves and other means'. The same accusations were made against the adventurers who expropriated land after the Cromwellian invasion, and half-hearted attempts were made to control them. But these efforts almost always failed. Even the forests of Glenkonkeyne and Killetra, which had been explicitly reserved for the Crown, were exploited by the London Companies with impunity.

Rather than providing the hulls for English ships, Irish timber was used for other purposes. Coopering, tanning, iron-making, glassmaking and housebuilding all used considerable amounts of wood. The coming of the new English settlers inaugurated a sort of industrial revolution in Ireland. The price of Irish timber – whether in the form of planks or staves – was much lower than in England or continental Europe, creating an economic logic to harvest and export it. Timber and staves were shipped to Scotland, England, Holland, Spain, France, the Canary Islands and various southern European ports. By 1615 Ireland was sending thirty cargoes of staves each year to the Mediterranean, and in 1625 it was said that France and Spain casked all their wine in Irish wood. Planks were exported to England for construction: Irish oak was used to rebuild London after Great Fire of London in 1666.

Undoubtedly, the upheaval in land ownership during this period created incentives for short-term exploitation. Millions of hectares of land changed hands as Gaelic chiefs and Old English lords were punished for rebellions or for choosing the losing side in English civil wars. Catholic ownership of land was reduced to 60 per cent by 1641 and to 27 per cent by 1688. After the Williamite War, when Irish Catholics backed the losing English King James II against the Protestant usurper William III, there were more forfeitures, so that just 15 per cent of land was owned by Catholics by 1703. These confiscations were never straightforward. Large areas of land became embroiled in legal proceedings and competing claims, and some transfers were later reversed. With such unstable land tenure, it is no wonder that some of those in possession of the land opted for short-term profit rather than long-term stewardship and decided to liquidate the value of their forests rather than manage them sustainably. Many of the soldiers, adventurers and undertakers always had an eye on making a fortune and then returning to England – Walter Raleigh is one example. Catholic lords who somehow clung on to land, but saw the writing on the wall, often made this calculation too.

Yet there were other settlers, such as Richard Boyle, who were in it for the long haul. Not all were asset-strippers. Wood-using industries that wanted to survive also had an economic interest in sustainable woodland management. The English ecologist Oliver Rackham, who happily fit the description of contrarian and brilliant Cambridge academic, even went so far as to argue that the only extensive Irish woodlands to survive to the nineteenth century were those that had ironworks nearby as this placed a value on the trees and ensured their protection. The best charcoal is produced from coppiced oak around twenty-five years old. If sustainable coppicing is practised and the supply of wood is in equilibrium with demand, ironworks and forests can co-exist. This was, in fact, what Boyle practised on his estates. At no stage in the thirty-year history of his iron-smelting business did he run out of timber for charcoal.

Oliver Rackham also pointed out that even if timber was harvested for short-term profit, trees would grow back in some form unless there was a more fundamental change in land use. 'When a wood disappears,' he wrote, 'one should not ask "Why was it cut down?" – for all old woods have been cut down from time to time – but "Why did it not grow again?"' What really drove the reduction of woodland cover during this period was increased demand for agricultural land. The three 'C's of conquest, colonisation and commerce led to an intensification of agricultural production as rural Ireland was integrated into a market economy at home and abroad. 'Land', according to Professor Raymond Gillespie of Maynooth University, 'was shaped into an estate system that created a framework for economic development through exploiting natural resources and building market structures.' It was like the Anglo-Norman manorial system on steroids. Low-productivity woodland was converted to higher-value agricultural use. Wood for staves, tanning, shipbuilding, housebuilding or iron-making was often a by-product of this land-conversion process. This was reflected in the financial accounts of Richard Boyle: he never made much money from his ironworks, but instead derived his considerable wealth from his land dealings.

Demand for agricultural products was partly driven by a growing and urbanising domestic population. Between 1550 and 1641 the population of Ireland doubled from roughly 1 million to 2.1 million, the highest rate of increase in contemporary Europe. The destructive wars of the 1640s led to a population collapse, but by the end of the seventeenth century numbers had exceeded 2 million again. By this time around 27 per cent of the population was of immigrant stock, descendants of the 350,000 people – English, Scottish, Welsh and French Huguenots – who migrated to Ireland over the previous 150 years. There was also rapid urbanisation during this period. Ireland was transformed from a predominantly rural society, with towns of Viking and Anglo-Norman origin confined to coastal or riverine locations, to one where market towns could be found in every part of the country. Dublin expanded from around 5,000 people in 1600 to 62,000 souls by 1703, by which time it was the largest city in the British Isles after London.

As well as filling plates at home, Irish farms became a source of food for England and its fledgling empire. Live cattle were the mainstay of the economy in the first half of the seventeenth century. Cattle prices in England had increased by more than a half since the 1590s, offering a booming market. The Irish live cattle trade to Chester grew from nothing in the late 1500s to 15,000 beasts a year in the late 1630s. Sheep farming also boomed: export figures for wool reached 6,666 stone per year by 1639, up from less than 200 in the 1580s. Cattle and sheep needed grass, which spurred much clearing of scrub woodland.

Because of pressure from British farmers, the government banned live cattle exports from Ireland through the English Cattle Acts of 1663 and 1667. Rather than killing exports, this just led to exports in a different form – the provisioning trade. This involved the export of beef, butter, tallow, fish and pork in barrels to England, France and the Low Countries and, more importantly, to the emerging English colonies in the Caribbean and North America. Ireland was part of a transatlantic economic system (and sent plenty of migrants to these colonies too). It was centred on Cork city, which drew in produce from the rich farmland of south Munster and got rich on the back of trade.

The growth of the provisioning trade had an impact on the flow of Irish timber. Irish exports of staves, planks and other timber products dropped sharply in the late seventeenth century. This timber was now kept at home, where it was used to manufacture the barrels needed to transport provisions. We can calculate the volume of timber needed to make these barrels, and it turns out to be almost exactly the same as the earlier exports of staves. In other words, timber was harvested from Irish woodland at the same rate as before but, instead of being sent to Spain or France to cask wine, it was used at home to make barrels for Irish produce. It was still a one-way flow, as once the barrels left they rarely found their way back. There was also more pressure on Irish forests via the leather-tanning industry, which now had many more cattle hides to process at home. The number of exported tanned hides rose from 106,300 in 1665 to 217,000 in 1669. Tanners could wreak havoc in a forest by stripping bark, which usually killed the tree.

The end result was the same: forests continued to be cleared to make way for agriculture and to produce the timber to support the agricultural economy. There is no doubt that the amount of woodland cover declined sharply between the start of the Tudor conquest and the final assertion of Protestant English control after the Williamite wars of the 1680s. We can read it in the tree rings. The average age of oak trees used in construction declined rapidly from a peak of almost 130 years in the mid-seventeenth century to just thirty-five years by the early part of the eighteenth century. This suggests that large, old-growth specimens were no longer available. Builders were forced to use younger trees grown in coppice systems – something that continued through the rest of the eighteenth century. There were simply no old oak trees left.

The idea of lost woodlands also began to appear in written texts at this time. The first study in the English language that can legitimately claim the title of regional natural history was written about Ireland by Gerard Boate, a Dutchman who located to London in 1630 and become the Royal Physician. His *Ireland's Natural History*, published in 1652 and dedicated to Cromwell, contrasted the lack of Irish woodlands at that time with the abundance described by Gerald of Wales five centuries before. He explained how the English had recently cleared the forests to increase the 'scope of profitable lands' and to produce charcoal for iron-smelting. He noted 'you may travel whole days long without seeing any woods or trees' except a few around 'gentlemen's houses'. He also warned that wood for fuel and timber was becoming scarce.

The clearest evidence of the dwindling wood resource is the fact that Ireland became a net timber importer. This was a major turning point in the history of Irish forestry. Softwoods were imported from the Baltics for construction in the second half of the seventeenth century. By the start of the eighteenth century Ireland was importing oak and bark to serve the needs of coopers and tanners. The iron furnaces spluttered and were shut down as local sources of charcoal dried up. By 1711 the value of Irish timber exports was £600, whereas the value of timber imports was a whopping £11,000. The flow of timber had reversed.

THE OLD NATIONALIST IMAGE of a Gaelic sylvan paradise destroyed in little more than a century by rapacious English invaders is a simplification. The forests were never as dense in the first place: the actions of the inhabitants of Ireland over millennia had already created a mostly open landscape. The woods that were left were not all high-canopy oak forests, but had large areas of scrub hazel and birch that were used for grazing and supplies. Yet, there had been large woodlands – by modern standards – in the more remote areas of the country, and these largely disappeared over the course of the seventeenth century. Moreover, Ireland had evolved from a position of relative abundance as a net exporter of timber to one of chronic deficit and a net importer of timber. Reversing this would be a preoccupation of the Anglo-Irish Ascendancy who emerged from this period in firm control of the land.

THE TWO IRELANDS

EDMUND BURKE WAS ONE of the intellectual titans of Ascendancy Ireland. Born in Dublin in 1729, he moved to London as a young man and forged a luminous career as a writer, thinker and politician. Now known as the father of modern British conservatism, he made his name as a defender of the rights of American colonists, as a stern critic of the French Revolution and as a campaigner against abuses of imperial power in India. He was also well placed to comment on Irish affairs. The son of a Protestant lawyer, he studied at the Protestant university, Trinity College Dublin (where he founded the 'Club', the first university debating society in the British Isles). But his mother was a Catholic, he was raised for a time by her family in the Blackwater valley in County Cork, and he was a lifelong supporter of Catholic emancipation. In his letters Burke bemoaned the division between Protestant and Catholics that cursed his homeland. They were 'not only separate nations, but separate species', living 'without common interest, sympathy or connection'. He wrote that there were 'thousands in Ireland who have never conversed with a Roman Catholic in their whole lives, unless they happened to talk to their gardener's workmen, or to ask their way, when they had lost it in their sports'. It was 'impossible that such a state of things ... must not produce alienation on the one side and pride and insolence on the other'.

The two nations – even species – Burke refers to was the dominant feature of Ireland during this period. Victory in the Williamite War in 1691 secured the Protestant Ascendancy for the next 200 years. Protestant planters owned most of the land, controlled government and maintained armed forces strong enough to repress any rebellions. Between 1695 and 1709 a succession of 'popery laws' or 'penal laws' were passed to limit Catholic political and civil rights. Catholics were an underclass until the middle of the nineteenth century.

The two Irelands had very different attitudes towards trees. There was a curious role reversal from the earlier period of conquest and colonisation. Before, English planters saw it as their civilising duty to clear forest 'wastes' wherever they could. Now, they became protectors of woodlands and proud creators of tree plantations. Then, Gaelic bards lamented the destruction of sacred trees and sylvan refuges. Now, the Irish peasant farmer saw forestry as an alien import and a competitor for precious land. Trees became loaded with political and cultural meaning, in a uniquely Irish way. So long as the Protestant Big Houses held a firm grip on the land, there would be a brief increase in forest cover in Ireland. But when the old order finally crumbled, and small farmers took back ownership of the land, trees would come tumbling down again.

TREE-PLANTING AND SUSTAINABLE forest management go hand in hand with secure land tenure and a long-term perspective. You are unlikely to plant a tree that will take fifty or a hundred years to reach maturity unless you believe that you, your children or your grandchildren will be in a position to benefit from it. Victory in the Williamite War gave the Protestant Ascendancy that security. It turned landlords into asset protectors. They set about investing in the land and reshaping it, in ways that are still evident today.

The planting and preservation of trees took on a cultural significance, as the ruling class continued their mission of improving and ordering the Irish landscape. The Big House, surrounded by its demesne, was at the heart of this project. Landlords sought to beautify their estates by creating lawns, ponds, avenues, orchards and forest plantations. In the late 1600s and early 1700s, the fashion was for geometry. Estate owners installed a checkerboard pattern of tree-lined avenues, canals and square blocks of trees in the French style, imitating the gardens of Versailles. By the middle of the eighteenth century a new romantic vision of man's relationship with nature had taken hold. The old geometric layouts were replaced by 'natural' parkland. Estate designers sought to enchant the observer with the picturesque and the sublime. Smooth, open meadows were dotted with clumps of trees; streams were re-routed to form lakes; animals grazed peacefully in front of romantic ruins, temples and pavilions. The demesnes were surrounded by great stone walls and belts of woodland to ensure privacy and to keep the ragged poor out of sight and mind. By the middle of the nineteenth century, parkland occupied around 325,000 hectares, or 4 per cent of Ireland, and over 7,000 houses featured ornamental or pleasure landscapes of four hectares or more. These demesnes remain one of the prominent man-made features of the landscape in Ireland today.

The other prominent feature of the landscape that dates to this period is the hedgerow. Outside the demesne walls, the land was enclosed and divided into an irregular grid of fields, each usually smaller than a couple of hectares. The enclosure movement arrived with force from England in the 1700s. Tenants were required to erect earthen banks and to plant trees around their holdings. Hedgerows and walls had a practical purpose: they helped control grazing animals and provided shelter on a windy isle. But they also delineated property boundaries and were seen as an essential element of a tidy Anglicised landscape.

Most of these hedgerows still survive today. If you fly into Dublin or Cork airport, and you are lucky enough to see between the clouds, one of the most striking features is the infinite network of hedgerows dividing the land below into pockets of green. Lined up end to end, the hedgerows of Ireland would stretch over half a million kilometres. They cover around 4 per cent of the country's surface. As a linear refuge for biodiversity, they are home to dozens of species of plants, birds, insects and mammals.

The hawthorn tree (*Crataegus monogyna*), a woody member of the rose family, forms the spine of most Irish hedgerows: it grows almost anywhere, has thorns to repel livestock, and is oblivious to hardship. But other species have found a home too. In the limestone lowlands, hedges are a mixture of hawthorn with hazel, ash, spindle (*Euonymus europaeus*) and guelder-rose (*Viburnum opulus*). The uplands are dominated by hawthorn with gorse or blackthorn (*Prunus spinosa*). At the wetter edges, scrub willows flourish; on the higher banks there are more than eighty forms of brambles or blackberry bushes (*Rubus fruticosus*); while the Atlantic seaboard is enlivened by the intense colour of fuchsia (*Fuchsia magellanica*), an import from Chile. Hedgerows are often cut low but occasionally ash or other trees break through and establish themselves as sentinels on the skyline. We tend not to think of hedgerows as a refuge for trees, but for 200 years there was a greater area of Ireland under hedgerow than under true woodland.

The spread of hedgerows and the development of the great estates were part of the cultural landscaping of Ireland. Trees were planted for ornament or shelter. But there was also an economic imperative for tree-planting: timber was increasingly scarce in Ireland.

This was a problem, because wood was still crucial to everyday life and industry. It was needed for building, tanning, fuel and storage. Before the advent of steel and plastic, liquids were for the most part kept in wooden buckets, casks and mugs. At the end of the eighteenth century brewers and distillers alone needed around 20,000 tonnes of wood annually to produce new casks. The provisioning trade used 380,000 casks each year to send beef, pork and butter abroad – equivalent to 7,000 tonnes of wood. Scarcity showed up in rising timber prices. In the early 1600s, for example, standing oak trees in County Cork were sold for just one shilling each. In 1731 oaks in the Shillelagh forest in Wicklow were valued at £3 17s each, and by 1780 the few remaining trees were worth £13 10s each. Indigenous timber supplies were so limited that the recovery of bog wood – semi-fossilised pine and oak that grew thousands of years before – was a profitable activity. The rural population turned increasingly to turf for energy, peeling back the bogs that smothered those earlier forests.

In 1703 the Irish parliament passed an act to encourage the 'Importation of Iron and Staves' as a means of counteracting the 'destruction of the woods' and the 'great scarcity of all sorts of Timber in this Kingdom'. Timber exports out of Ireland dried up almost completely. Timber imports grew in value from £11,000 in 1711 to £187,400 per year by 1790. These imports consisted of pine or fir boards, oak staves and raw logs, mostly from the Baltic region, although North America would later become the dominant supplier.

By the 1790s Ireland was also importing 8,000 tonnes of British oak bark for its tanning industry each year. The volume was so great that its impact could be seen in English woods. In 1796 William Marshall published a survey of the forests of south-west England. Noting that the 'market for bark is Ireland', he warned that the 'exorbitant price' of this commodity threatened national security through the 'annihilation' of trees that would otherwise have been preserved for the Royal Navy. This was the old narrative turned on its head. Instead of the Royal Navy clearing Irish oak to build its ships, Irish tanners were now destroying English oak destined for the Royal Navy! It was still not enough, though, and during the 1700s most Irish hides were exported raw and finished abroad because of the lack of oak bark at home.

The scarcity of timber and dependence on imports became a matter of concern for gentlemen committed to the improvement of the country. One of the most eloquent was Jonathan Swift, Dean of St Patrick's Cathedral and author of famous satirical works such as *Gulliver's Travels*. In an excoriating attack on Irish political affairs, the *Drapier's Letters* published in 1724 and 1725, he lamented how his fellow countrymen neglected to plant new trees, failed to enclose their woodlands and cut down their woods prematurely for quick profit. 'There is not another example in Europe,' he wrote, 'of such a prodigious quantity of excellent timber cut down in so short a time, with so little advantage to the country, either in shipping or building.'

Such grumblings led to political action. The Irish House of Commons passed an Act for Planting and Preserving Timber Trees and Woods in 1698. The preamble stated that 'the timber is utterly destroyed' because of 'the late rebellion in this kingdom and the several ironworks formerly'. Between 1698 and 1791 seventeen parliamentary acts were directed at conserving and increasing the area under woodland in Ireland. The earlier acts levied fines for illegal felling of trees, forbade the keeping of goats on mountain lands and tried to limit the use of wattling in the walls of buildings – which had been a standard construction method in Gaelic Ireland. They required landowners, tenants and ironmasters to plant a minimum number of trees, depending on the size of their holdings, and to enclose these plantations with a stout wall, hedge or fence. They even banned the making of maypoles, a tradition that never really caught on in Ireland. Later acts tried to encourage planting by giving tenants more security over trees they planted and giving them rights to harvest timber from these trees. Some acts set targets for tree-planting – 230,000 trees per annum was the target in 1698 and again in 1705 – something that will be familiar to followers of contemporary Irish forest policy. Like today's targets, they were usually missed.

Rather than relying on the government, the gentlemen of Ireland decided to take matters into their own hands. In 1731 the Dublin Society for Improving Husbandry, Manufactures and Other Useful Arts was formed. (It would later be known as the Royal Dublin Society, or the RDS.) According to one historian, it became 'the principal agent of economic development in the country'. In 1738 Reverend Samuel Madden, a close friend of Jonathan Swift and a leading light of the Dublin Society, published his *Reflections and resolutions proper for the gentlemen of Ireland*. He wrote of a nation where nearly a quarter of the 'profitable land was under vast forests' less than a century before. It was therefore 'strange' that 'we should now be reduced to a necessity of planting, or lie under an increasing expense of £40,000 per annum' for imported timber. He urged 'the gentlemen of Ireland' to operate on a more considerable scale, undertaking 'large and noble plantations'.

The next year the Dublin Society raised a charitable fund and began offering premiums or awards to landowners who planted new woodlands. The scheme grew and from 1761 was funded by the Irish parliament. The society elaborated a set of silvicultural requirements to be eligible for support. For example, the 1783 scheme required landowners to plant at least four hectares, to use a list of approved species, and to fence and maintain the plantation for ten years. It also defined a minimum stocking rate of 3,000 saplings per hectare. (As we shall see, this is very similar to modern-day afforestation schemes.) Overall, between 1766 and 1806 the society paid out £19,000 in premiums on 55 million trees, or around 3½ pennies per tree. This sounds like a lot of trees, but the total area of new woodland was probably less than 10,000 hectares, which equates to a planting rate of less than 250 hectares per year over forty years. After the Act of Union in 1801, parliament reduced funding to the Dublin Society and the forestry premiums ended a few years later.

The forestry schemes of the Dublin Society give a glimpse into the enormous diversity of trees that were planted on the great estates during the eighteenth and nineteenth centuries. In 1744 the society specified that three native species (oak, ash and elm) and three European species (beech, walnut and chestnut) could be planted to avail of the premiums. By the 1780s the list of approved species stretched to scarlet maple, cedar of Lebanon, American elm and birch, the tulip tree, black larch, Newfoundland spruce and two-thorned acacia – a veritable arboretum. The gentlemen foresters of Ireland scoured every corner of the empire for exotic species to experiment with. There was a competitive element to this: unusual trees were a status symbol, a towering ornament to embellish an estate.

The Ascendancy planters were the ultimate vector for the return of tree species to Ireland that had struggled to make it back after the end of the last glaciation. Some of these species had been introduced by previous settlers over thousands of years, but they had never been planted on such a large scale. Starting in the 1700s, a whole new range of species muscled their way into the canopy, changing the composition of Irish woodlands forever.

Sycamore (*Acer pseudoplatanus*), a member of the maple family, was widely planted at this time, and it flourished in its new home. A native of central, eastern and southern Europe, it grows quickly and can form a huge tree, although its lifespan is surprisingly short, usually less than 200 years. It produces vast quantities of winged seeds that twirl through the air like a helicopter when they fall in autumn. Sycamore wood was used for making churns, butter bowls and other utensils for dairying, as it had a fine texture and did not impart any taste to the milk. Sycamore is now the second most common large hedgerow tree in Ireland (after ash) and has established itself in many 'natural' woodlands, having seeded in from nearby estates.

Beech is another naturalised tree that was widely planted in the 1700s and 1800s. The bark is smooth, thin and grey, and it has broad, oval leaves with wavy edges. Beech is the climax forest species across millions of hectares in central Europe, growing to a tremendous height. Beech employs a patient strategy to win out over other species, establishing itself in the shade under an existing canopy and then growing up over many decades in the gaps until it overtakes the other species and dominates the forest. Beech woods are quiet and cathedral-like, enveloped in deep shade and carpeted with a dense mat of fallen leaves, which combine to prevent many plants from growing on the floor. Beech was well suited to the drier soils of eastern Ireland and can be found in old estates such as Powerscourt and in many city parks.

Children will be thankful that the horse chestnut (*Aesculus hippocastanum*) was a favourite of Ascendancy planters at this time. Originally native to the Balkans, it was introduced to England in the 1500s and was brought from there to Ireland. It forms a large tree, up to 40 metres high, and is found in parks, gardens and streets. Its most distinctive feature is its large seed, the smooth mahogany-coloured 'conker'. Conkers burst from a spiky green husk in the autumn, and are eagerly collected by children, who thread string through the centre and smash the conkers together to find out which lasts the longest. No one is sure how the horse chestnut got its name. One explanation is that conkers were ground up and fed to horses to relieve them of coughs. The other is that its leaf scars, left on twigs after the leaves fall off, have a horseshoe shape.

These new deciduous species blended in naturally with the remaining patches of ancient Irish woodland. The same cannot be said for a class of tree that estate owners began planting in the eighteenth century – the conifers. Conifers have needles rather than leaves, and bear their seeds in cones rather than in fruits or capsules. The difference is most stark in the winter, when most native Irish trees lose their leaves and stand naked in the low sunlight. In contrast, the conifers, or evergreens, bear their needles throughout the season, standing out as patches of dark green on the landscape. Coniferous softwoods were (and still are) an attractive tree for planters, because they grow fast, tall and straight, and produce long planks and boards that can be easily worked for construction. Ireland was importing large quantities of softwood from the Baltics, so it made sense to try to develop a homegrown resource.

One of the first conifers to be planted widely was Scots pine. This native Irish tree had been abundant for thousands of years before largely dying out, apart from a few relict stands such as at Rockforest in County Clare. Most Scots pine trees today trace their origins to seed brought in from Scotland in the 1700s and

1800s and planted on Irish estates. Another conifer introduced at this time was Norway spruce (*Picea abies*). The natural range of Norway spruce is northern Europe, from Scandinavia east to Russia, as well as the mountains of central Europe. It is a fast-growing evergreen that grows tall and straight with a triangular appearance – it was often used as a Christmas tree. One other conifer traditionally used as a Christmas tree, also introduced to Ireland, was European silver fir (*Abies alba*). It grows in mountains throughout Europe and has distinctive silvery-white needles. But the term 'fir' was often used to refer to Scots pine, Norway spruce and silver fir in the eighteenth century, so it is not always clear from written records which species of evergreen conifer was planted. Often it was a mix. As early as the 1680s, Lord Granard was planting 'large groves of fir of all sorts' around Castle Forbes in County Longford.

Perhaps the most fashionable and widely planted conifer in the eighteenth century was European larch (*Larix decidua*). It is unusual in that it sheds its needles each winter, making it a deciduous conifer. Bright green foliage in spring changes to a darker shade of green through summer before turning a golden yellow in autumn. The European larch is native to the Alps and the Carpathian range of central Europe. It is very cold-tolerant, surviving in temperatures as low as minus 50°C, and it is often the highest tree in these mountain ranges. It only grows well on well-drained soils and is a light-demanding species, so it is a good pioneer species on bare ground and is sometimes used as a 'nurse' tree before introducing more demanding species. The tree produces a hard timber that is especially suitable for boat-building: for example, it is used to build the hulls of traditional Galway hookers. Larch was one of the first trees to be introduced to Britain for its timber, and it became a popular choice for timber plantations in Ireland.

Conifer plantations were perceived very differently to now. As well as being a mark of Enlightenment progress, they were seen as beautiful in their own right. In 1738 William Ellis

published *The timber-tree improved*, in which he championed
the new conifer species for 'their Uniformity and Beauty; their
perpetual Verdure; their Sweetness; their Fruitfulness, affording
Seeds, Gum, Fuel and Timber, of all other Woods, the most
useful and easy to work'. Twelve years later a physician from
Waterford city, Charles Smith, published a treatise on County
Cork that was full of praise for conifers. He described the
plantations of Lord Egmont around Burton House, planted
with 'ash, elm, oak, and large quantities of fir; than which last,
no timber tree, in the winter season, affords more beauty to a
landscape'. There was a different landscape aesthetic at this time.

What was the overall impact of estate tree-planting on Irish
forest cover? We are fortunate that, for the first time, there is
census data from the late eighteenth century showing how land
was used in Ireland. Starting in the 1830s the country was also
mapped in detail by the Ordnance Survey. The census returns
indicate that 42,500 hectares had been planted with trees by
1791. This was 0.5 per cent of Ireland's total land area, a modest
effort. Indeed, when the Registrar of the Dublin Society looked
back at the history of this period a century later, he concluded
that the premium scheme 'seems to have done a great deal
indirectly, and by force of example', but 'it did not, however,
lead to the creation of a single plantation on a really large scale'.

The Irish census only records information on *planted* trees. It
has little to say on natural, ancient woodlands. They are lumped
into a category of land termed 'Uncultivated', which includes
bogs and mountains, and altogether accounted for 30 per
cent of the total landmass. There is plenty of evidence that the
remaining patches of natural woodlands continued to be cleared
in the 1700s. The historian Eileen McCracken identified 400
advertisements for woods for sale in Irish newspapers between
1730 and 1779, usually mature woods ready for harvesting. Irish
landowners tended to sell standing trees to timber merchants,
who would then send their own crews in to harvest. These were
likely to be natural woodlands rather than the plantations that

were only being established. McCracken estimates that 22,000 hectares were sold during this fifty-year period, or an average of 440 hectares per year, which is more than the planting rate under the Dublin Society premiums.

Some of the last great ancient woodlands of Ireland were whittled down during this period. One example is Shillelagh. Nestled in a verdant corner of southern County Wicklow, it was one of the most famous and productive Irish forests in the 1600s, supplying oak timber for the roof of Westminster Hall in London and sending a steady stream of staves down the river to Enniscorthy. Documents show that woods covered 2,266 hectares in 1656. The land was owned by the Watson-Wentworth family whose most famous scion, the second marquis of Rockingham, was British prime minister twice. By the early 1700s a survey found there were 485 hectares of woodland remaining, with 2,000 mature oaks that were 'the Glory and ornament of the kingdom of Ireland'. But ironworks moved in to the area and the great oaks were felled for timber: a mere thirty-eight lasted to 1780. Rather than renewing the woods, the Watson-Wentworth family converted much of the land to agriculture: estate records show that income from woodlands fell from one-quarter of agricultural rents in the 1740s to just 4 per cent at the end of the century. The last remnant of this once mighty oak forest is Tomnafinnoge Wood, a 66-hectare conservation area with peaceful trails that wind alongside a shaded river. Now owned by the state, it was almost cut down in the 1980s but was saved by a public campaign that attracted the support of then taoiseach Charles Haughey.

The clearance of native woodlands was part of a final de-wilding of Ireland that took place in the eighteenth century. The last recorded wolf was hunted down and killed in 1786 near Mount Leinster in County Carlow. The capercaillie, a giant woodland grouse with an elaborate mating ritual, disappeared around the same time. Even the red squirrel, an iconic forest species, was considered extinct in Ireland by the end of the eighteenth century and had to be re-introduced from England in the mid-1800s. This is a clear sign of how scarce woodland had become. Indeed, when the ecologist Oliver Rackham compared the Ordnance Survey maps of the 1830s with the Cromwellian surveys of the 1650s he concluded that only one-tenth of the original woodland had survived.

The total amount of woodlands continued to decline because the rate of clearing of natural woodland was higher than the rate of new tree-planting. Forest cover – including both plantations and natural woodlands – probably dropped to between 1 per cent and 1.5 per cent of the land area by 1790. The treeless state of Ireland was often commented on by visitors at this time. When Arthur Young, a leading English agriculturalist, toured Ireland between 1776 and 1779, he wrote that 'the greatest part of the kingdom exhibits a naked, bleak, dreary view for want of wood, which has been destroyed for a century past, with the most thoughtless prodigality'.

The situation was reversed for a time in the nineteenth century. The most intensive wave of tree-planting came in the early 1800s, propelled by legislation, new methods of forestry and a boom in demand for timber, especially oak, caused by the Napoleonic Wars. By 1841 the area under tree plantations had increased to 139,860 hectares. This was a tripling in fifty years, or a rate of planting of around 1,900 hectares per year. (In addition, through some heroic statistical efforts, the census-takers calculated there were 11,686 hectares of orchards and 45,760 hectares of 'detached trees' in hedgerows and ornamental plantings by 1841.) Over the next forty years the area under plantations remained fairly constant. The 1881 census records a figure of 132,766 hectares. By this time, forest cover in Ireland – combining plantations and the small area of natural woodland – had recovered to around 2 per cent. This was still very low by European standards: the total area under forestry at this time was 4.5 per cent in Scotland and 5.2 per cent in England, and as much as 25 per cent in Germany. But it was the first time there had been a significant increase in Irish woodlands since the aftermath of the Black Death in the Middle Ages, so it is notable all the same.

THERE WERE MORE WOODLANDS in Ireland but they were very different from before. Most were planted by human hand rather than put there by nature. They were more exotic in species, and increasingly tilted towards conifers. Their location reflected social, economic and cultural realities; they were mostly found on the estates of the Protestant Ascendancy, in and around their demesnes. The peculiar features of Irish forestry were remarked upon by a number of travellers in the nineteenth century. In 1837 Samuel Lewis published *A Topographical Dictionary of Ireland*, the most detailed survey of the country in the years before the Great Famine. 'The country', he wrote, 'is extremely deficient in timber.' Its 'ancient forests' had 'long since been cleared away' and 'not until lately' had 'any general or enlarged exertions been made to reinvest the country with this useful and beautiful appendage'. 'The only plantations', he continued, 'are in the neighbourhood of the mansions of the nobility and gentry.' A French traveller in the late 1880s made a similar observation on the association of trees with the Big House. 'They are seen only in private parks,' he wrote. 'The tree has become a lordly ensign. Wherever one sees it one may be certain the landlord's mansion is not far.'

At this time there was a huge gulf between the landlord class and the masses below. The majority of the Catholic Irish lived a precarious existence. Half the population consisted of cottiers or labourers, who lived in single-room mud cabins and only had access to small plots of land, usually less than one hectare, on which to grow crops and graze a cow. Their leases were insecure, land could be taken away at short notice, and they relied on work on the estates that could easily dry up. They were at the mercy of calamities that could, with the slightest touch, tip them into destitution or starvation. The same visitors to Ireland who commented on its treeless landscape noted its extreme poverty compared to other European countries.

Wood was still a precious resource for these impoverished farmers and labourers. It provided fuel, shelter and materials for the manufacture of implements. Irish men and women continued to carry their axes into the ever-dwindling ancient woodlands to harvest what they needed. This brought them into conflict with the major landowners – the Protestant Ascendancy – who now positioned themselves as protectors of trees against the short-sighted actions of the peasantry. Arthur Young, the visiting English agriculturalist, pointed the finger in this direction in the 1770s. He wanted to know why there were so few trees when the landscape offered 'evident signs' of having been 'once wood or at least well wooded'. The 'gentlemen' he spoke to ascribed the 'destruction' of Ireland's woods to the 'common people … who, they say, have an aversion to a tree; at the earliest age they steal it for a walking-stick; afterwards for a spade handle; later for a car shaft; and later still for a cabin rafter'. This was because of inadequate supervision and encouragement of the 'common people'. This conflict over resources was exacerbated by the new tree plantations that the Ascendancy class were establishing. The Catholic rural poor had a desperate hunger for land at this time; land planted to trees was land that could have been cultivated to feed a family.

The tensions and divisions in rural Ireland led to recurring waves of agrarian violence, as Whiteboys and other secret tenant groups formed to inflict revenge on unsympathetic landowners. Eighteenth-century estate papers reveal that forests and orchards were sometimes attacked. For example, Philip Cosby offered a reward of 20 guineas – equivalent to a full year's wages for a butler – for information after some oaks were cut in his woods at Stradbally in County Laois. The Duke of Leinster, when trees were felled on his estate at Carton in County Kildare, announced that he had set man-traps in his plantations. After the Whiteboys launched a night assault on Dromana in County Waterford, seat of Earl Grandison, a correspondent wrote to William Pitt that they 'levelled a part of the Deer Park Wall erected by his Lordship around a large Forestry on his Demesne'. The Whiteboys claimed that it had been common land a hundred years before.

The Whiteboys' justification hints at the other side to this story. Gaelic writers lambasted landlords as 'foreign devils' who whipped poor Irish neighbours for entering property to collect wood for fuel, denying people customary rights that they had enjoyed for centuries. Folk memories survive in ballads that tell how Irish youths were hanged by cruel landlords for stealing branches from demesnes. This was a bitter conflict over control of the land.

The political and religious tensions of this era led to another ill-fated rebellion in 1798. It started with the non-sectarian, democratic and republican ideals of the United Irishmen, who planted Trees of Liberty to symbolise their cause. As the United Irishmen plotted rebellion, they tasked blacksmiths with manufacturing pike blades and sought out young ash trees to cut pike handles. This was done surreptitiously, but once a French army landed and the insurrection broke out, many demesnes were raided and trees cut to make pikes. The rebellion descended into an orgy of sectarian violence and massacres on both sides that left 30,000 dead. The Liberty Tree was replaced by the Hanging Tree, as hundreds of defeated rebels were executed by Crown forces. For years, rumours of ash trees being cut down at night were taken as warning of further rebellion.

Despite this bloody episode, and all the poverty and hatred, the Irish population grew at a spectacular rate during this period, faster than anywhere else in Europe. The number of people more than tripled from 1730 to 1845, increasing from 2.5 million to more than 8 million. Ireland had a population density of 270 people per square kilometre by the early 1840s, second only to Belgium. This demographic explosion was made possible by one nutritious plant – the potato. Imported from the Americas by English colonists, it grew well in Ireland's poor soils and damp conditions. By 1845 two-fifths of the population depended chiefly on potatoes for survival. Smallholders pushed this crop into every corner of Ireland. They cultivated the barren soils of the western seaboard and the boggy sides of hills by creating 'lazybeds', long ridges about a spade's width fertilised by seaweed, manure or shell sand. You can see traces of these abandoned fields on wind-beaten hillsides across Ireland today.

The expansion of agriculture, propelled by population growth, led to further clearing of trees in Ireland, especially scrubby woodland. Observers noted how large farmers took the best land and planted it to grass for cattle, pushing the poor farmers into the mountains, bogs and wooded areas. Cottiers and labourers did the hard work of clearing shrubs and trees, cultivating the soil for potatoes and herding cattle and sheep into every nook and cranny of the landscape. In their desperate search for land, they were a powerful force in the de-wilding of Ireland.

The landscape of Ireland bifurcated. Inside the demesne walls the new plantations and old woodlands were protected by a landlord class that had the luxury of applying long-term silvicultural management. Outside, the remaining patches of woodland and scrub were cleared by a desperate peasantry, creating an almost treeless landscape. This hunger for land left no corner untouched. For example, it was once thought that the trees that can be found growing on remote islands in lakes throughout Ireland might have an unbroken connection with the primeval forests. They seem inaccessible to humans and protected from grazing animals. But every time scientists study these islands, they invariably find that they were cleared at some point in the last 300 years. People were prepared to row out by boat, taking their tools, cattle and sheep with them, to plant crops or graze livestock. Every inch of ground was utilised in some way.

We can see these forces at work on one of the most remote and inhospitable terrains for human settlement, a place where the trees never grew back – the Aran Islands. Scientists have reconstructed the woodland history of the islands by studying pollen samples in cores drilled from the largest lake on the islands, An Loch Mór. If Ireland is 'an island off an island off the coast of Europe', the Aran Islands off Ireland's Atlantic coast are one step beyond. But seeds had reached there as part of the postglacial tree migration, and the islands were once densely wooded. The original woodland consisted of pine, oak and elm, with hazel important as an undershrub. Trees were cleared for Neolithic farming, and again in the Iron and Bronze Ages, with plenty of lulls along the way leading to bursts of woodland regeneration. Farming activity really picked up in the medieval period and the landscape was gradually opened up over the following centuries. It was in the eighteenth and early nineteenth centuries that the last trees were cleared, as the population grew and the introduction of the potato led to an expansion of arable farming. The Aran Islands, once blanketed in woods, are now famous for having almost no trees at all.

The population explosion and reliance on a single crop led to the ultimate calamity in 1845: the Great Famine. The potato blight, a fungus called *Phytophthora infestans*, wiped out the staple crop of millions of people. An inept and half-hearted response by government and the wealthy turned a crisis into famine. More than 1 million people died and another million emigrated, reducing the Irish population by a quarter within a few years. Thanks to continuing emigration, the population kept falling for the rest of the century.

In previous eras population collapses had been followed by the regeneration of woodland, but this did not happen after the Great Famine. Instead, people were replaced by animals. The livestock sector underwent a massive expansion. This came about because of the further commercialisation of Irish agriculture and the further integration of Ireland into the British food economy. There was a pull factor at work (the industrialisation and urbanisation of Britain created a huge demand for food) and a push factor (the construction of railways opened up every corner of Ireland and allowed food to be moved quickly to British tables). The total number of cattle in Ireland grew from 1.8 million in 1841 to 3.8 million by 1861. Cattle were bred in the poor, boggy uplands and fattened in the limestone valleys of north Munster, inland Connacht and north Leinster. Smallholders were displaced by the grazier class, and the 'strong' farmers grew richer and intensified their effort to put as much of the country as possible under grass. The abandoned lazybeds and potato fields, rather than sprouting the saplings of trees, were trampled and grazed short, as Ireland renewed its Celtic love affair with the cow.

THE ENRICHMENT OF THE 'strong' farmers was accompanied by a revival of the Catholic Church, Catholic political power and nationalist sentiment under leaders such as Daniel O'Connell and Charles Stewart Parnell. They trained their sights on land redistribution as a political goal. This pressure paid off when the London government introduced a series of Land Acts at the end of the 1800s that forced the break-up of the Anglo-Irish estates. Millions of hectares of land were transferred to tenant farmers, who were financed by long-term government loans. In 1870 only 3 per cent of Irish farmers owned their land, while the rest were tenants. By 1929 the situation had been completely reversed, and 97 per cent of farmers owned their farms in freehold. A new class of peasant proprietors was born at the expense of the old Protestant Ascendancy.

Land reform had a destructive impact on Irish woodlands, setting back much of the progress that had been made over the previous century. Among the estate owners, it turned asset protectors into asset-strippers again. Landlords who had carefully nurtured their tree plantations decided to cash in before their land was sold to their tenants. English timber merchants flocked to the country to buy standing timber for export back to England. Eileen McCracken writes that they 'moved across the country from estate to estate like arboreal pests'. And because they foresaw the land slipping from their grasp, estate owners were reluctant to establish new plantations. A.C. Forbes, future director of forestry at the Department of Agriculture, commented that '1880 seemed to mark the almost total cessation of planting on private estates'.

At the same time, the former tenants who became landowners were motivated to fell any woodlands that came with their newly acquired land. They had an urgent need for capital, not least because their land purchase was financed with government loans that had to be serviced. They were intent on maximising near-term income through livestock or crops. There were also cultural factors at work. In the eyes of many small farmers, trees were associated with the Protestant Ascendancy and those close to them. The tree plantation, in particular, was seen as a foreign import. No culture of forestry or woodmanship survived among the rural community, unlike in other parts of Europe, partly because there were so few trees around. Farmers wanted to see grass or crops growing from hedgerow to hedgerow, and saw little point in putting farmland under trees, especially as these trees would take decades to produce a return.

By 1907 plantations and natural woodland covered just 119,400 hectares in Ireland, which was 1.4 per cent of the total land area. There was another pulse of deforestation during the First World War. German U-boats sank ships carrying imported timber to the United Kingdom, leading to severe shortages and price spikes. The British prime minister, David Lloyd George, later admitted they came closer to losing the war from lack of timber than from lack of food. An estimated 12,000 hectares of Irish woodland were felled to serve the war effort, and much of this area reverted to scrub. By the time of the birth of an independent Ireland in 1922, forest cover had been reduced to around 1.25 per cent. This was at least as low as the 1790s. Indeed, it probably represented the lowest amount of woodland on the island for 11,000 years. Ireland was not quite a country with no trees, but it was not far off.

Often, the only trees that survived were ones that were

considered sacred. The tradition of woodmanship may have died out, but the Celtic legacy of sacred trees lived on, along with the taboos that protected them from felling. The hawthorn, also known as whitethorn, benefited most from this protection. A lone hawthorn was associated with fairies, the otherworldly beings who lived in parallel with the normal world. Hawthorns growing around ancient monuments such as forts, wedge tombs or burial mounds were especially magical. For example, a German traveller in 1842 described a fairy mount in County Longford where no one would touch any wood growing on it. 'Young trees they will steal with very little remorse,' he wrote, 'but wood growing on one of these fairy mounts is almost always secure from their depredations.' Folk stories warned of the bad luck that would befall anyone who damaged these trees. Eyes were blinded by thorns, skulls split by flying axe heads, legs broken by falling branches, minds driven to insanity. As trees became scarcer, these trees stood out even more, causing them to be more venerated. You can still see these lone fairy trees standing in the middle of fields throughout Ireland, cautiously left by a farmer even though he has to drive his tractor around it every time.

THE BEGINNING OF THE twentieth century was the nadir of Irish forestry. The revolution in land ownership unleased a wave of woodland destruction, just as it had in the 1500s and 1600s. But a force was stirring that would reverse the tide of deforestation and set in motion tree-planting on a scale never seen before. Over the next 100 years Ireland would go from being a net importer to a net exporter of timber again. The age of reforestation was about to begin.

REFORESTING IRELAND

FOR MANY YEARS A staple of Irish school tours was a pilgrimage to Avondale House and Forest Park. This is an old 214-hectare estate perched on the foothills of the Wicklow Mountains and overlooking the picturesque Vale of Avoca. Children can marvel at the tallest trees in Ireland and exotic species such as giant sequoia from California and monkey puzzles from Chile. They can collect conkers, acorns, pine cones and the weird and wonderful fruits of dozens of species as they wander along the damp, shaded trails. Usually, the children ignore the memorials scattered around the park to long-dead Irish foresters who helped create the woodlands, yet Avondale is home to the most diverse, experimental and consequential collection of trees in the country. It has been called 'the cradle of Irish forestry'.

Like counting tree rings, the story of Avondale reflects many of the key chapters in the history of Irish forestry. Avondale House was built around 1779 by Samuel Hayes, the author of *A Practical Treatise on Planting; and The Management of Woods and Coppices* and an influential advocate of estate tree-planting in the late eighteenth century. He planted woodlands at Avondale, some remnants of which survive to this day. The estate was later owned by Charles Stewart Parnell, the nineteenth-century leader of nationalist Ireland who did so much to advance the land reforms that would lead to the break-up of the Anglo-Irish estates and, indirectly, the destruction of their woodlands. Avondale was not immune to the wave of tree-felling around the turn of the century: there was a sawmill in operation until 1904, by which time the stock of marketable timber was exhausted. A photograph of the house around this time shows a mostly bare landscape, dedicated to grazing, with just a few pockets of estate woodland.

In 1904 the government purchased the estate and opened Ireland's first state forestry school. Starting with eight apprentices in 1904, the new school trained an early cadre of foresters who went on to establish state forests around the country. Avondale was also used to trial new tree species: the foresters laid out 100 plots of an acre each and planted more than forty species, some of which have become the backbone of the Irish forest industry today. Avondale was later transferred to Coillte, the semi-state company that was formed in the 1980s to manage Ireland's public forests. A marvellous Centenary Trail was opened a few years ago, taking visitors through the experimental tree plots established more than a century before. Avondale is now being redeveloped in conjunction with a German leisure company that has invested €8 million to build a treetop walkway and a spiral viewing tower that rises above the forest canopy.

The opening of the Avondale forestry school was the beginning of a new chapter in the history of Irish woodlands. Henceforth, the state would play a much more powerful role in the protection of existing woodlands and the planting of new forests. Avondale also marked a shift towards 'scientific forestry' and a reliance on a new suite of fast-growing conifers imported from another continent. The new policies reversed the tide of deforestation that had washed over the island for thousands of years. During the twentieth century, trees were planted in Ireland on a scale never seen before.

THE FOUNDING OF THE Avondale forestry school was a response to public clamour over the rampant tree-felling that followed the Land Acts. A string of committees, societies and reports around the turn of the twentieth century called attention to the dire state of Irish woodlands. They advocated for a radical new forest policy as part of a general national revival. James Joyce parodied these nationalist concerns in one of the dialogues in his novel *Ulysses*, set on 16 June 1904, the same year that Avondale was established. 'As treeless as Portugal we'll be soon ... or Heligoland with its one tree if something is not done to reafforest the land,' says John Wise Nolan, one of the minor characters in the Cyclops chapter, holding forth in Barney Kiernan's public house. 'Larches, firs, all the trees of the conifer family are going fast.' 'Save them,' the citizen responds, 'Save the trees of Ireland for the future men of Ireland on the fair hills of Eire.'

There was a consensus that only government action could stop the destruction of the few remaining woodlands and achieve the planting of new forests on any scale. The rationale for government intervention was rooted in the long-term nature of forestry. It was difficult for a private landowner to invest in planting, to forego any agricultural income from the land and to wait the fifty or hundred years that it might take for the trees to generate an economic return. Only the state could take this long view. The Irish Minister for Lands and Agriculture, Patrick Hogan, articulated this logic in a Dáil debate in 1928:

> Unless the State takes a very active part in afforestation work nobody will do it. It is an enterprise which cannot be carried out except by a man who has plenty of capital and who is in a position to wait for a long time for his return. There are very few such people here.

In addition, forestry delivered spillover benefits for the broader community. A vibrant forestry sector created jobs in forest management, timber processing and auxiliary industries while supplying important goods to the local economy and displacing imports. These wider benefits did not accrue to individual landowners, but they were important to governments who saw forestry as a tool for economic and industrial development.

Advocates of a more forceful forest policy were also seduced by the idea of turning the wastelands of Ireland into something more valuable. This was a modern version of the settler obsession with 'improvement' in the 1600s and 1700s. There were hundreds of thousands of hectares of bog and rough grazing land, usually in the poorest areas of the country, that generated little profit from agriculture. It was thought that growing timber would generate a higher economic return while creating jobs. More fancifully, dreamers on a big scale believed that trees would improve the quality of the land, turning the bleak open landscapes of the west into a farming arcadia. For example, in 1884 British prime minister William Gladstone commissioned a study by a self-appointed forestry expert named Daniel Howitz. He produced a report titled *The Reafforesting of Waste Lands in Ireland*, in which he recommended planting 1.2 million hectares along the western seaboard to create a giant shelter belt. This would protect the land behind from Atlantic gales, transforming its productivity. Through an alchemical process, trees would turn muck into brass.

The first effort at state afforestation in Ireland had these objectives – and it was an utter fiasco. It took place on a wet, rocky hill at Knockboy in Connemara in the 1890s, the brainchild of a local priest, Father Thomas Flannery, who responded to signals from the government that there were funds available for afforestation by presenting 400 hectares of bogland in his parish. After some bureaucratic pass-the-parcel, the scheme ended up in the hands of the Congested District Board, which planted more than 2 million trees of thirty different species and invested £10,000 (around £820,000 in today's money) in an attempt to establish a forest. But within a few years almost all the trees were dead and the plan was aborted. A contemporary wrote that 'a rockier or more wind-swept spot than Knockboy may not be found in all Ireland, and had the trees found root in its sterile sheets of rock, or sustained the unchecked onslaught of Atlantic storms, the difficulties of forestry elsewhere in Ireland must have been small indeed'. If you visit today, all you will find are a few bent trees that were able to find refuge in shallow depressions with slightly better soils.

Knockboy was a false start. The true beginnings of state afforestation in Ireland can be found on the other side of the country at Avondale. The man put in charge of the new state forestry school – A.C. Forbes – would play a dominant role in

Irish forest policy until the 1930s. Forbes was described by a contemporary as 'an Englishman, who is, by the way, devilish uninteresting outside the subject of forestry'. He had authored a number of books on silviculture and lectured at Newcastle University before moving to Ireland. On taking over at Avondale he decided to develop an experimental forest garden to determine which trees would grow best in Irish conditions. He laid out the 100 plots that we can still see today. If Avondale is the 'cradle', Forbes has been called 'the father of Irish forestry'. He defined the philosophy and work ethic of the early state Forest Service, and his views would shape Irish forests for generations to come.

Soon after his arrival in Ireland, Forbes was asked to provide technical input to what became the foundational policy document for Irish afforestation. In 1907 Dublin Castle appointed a Departmental Committee to make recommendations on 'the financial and other provisions necessary for a comprehensive scheme of afforestation'. This committee held fifteen meetings, examined forty-eight witnesses, compiled an enormous amount of statistics and, the following year, published a 500-page report that is still a mine of useful information on all things related to Irish forestry. The committee argued that 'a national scheme of afforestation cannot be undertaken by private individuals', but required the intervention of the state, 'a proprietor who never dies'. They concluded that an area of at least 1 million acres (or 405,000 hectares) of woodland was 'essential for the agricultural and industrial requirements of the country'. They recommended that the state set a target of acquiring 80,000 hectares for planting over the following ten years.

The report made an impact. The Westminster parliament voted £6,000 for Irish afforestation and the Department of Agriculture started to acquire sites in Cork, Laois and Wicklow. By 1922 Forbes had overseen the planting of 1,295 hectares of land and the purchase of 1,456 hectares of existing woodland. This was far below the target set in 1907 but a start nonetheless. Ireland was a pioneer in state forestry within the United

Kingdom; it was only after the First World War that a Forestry Commission was established to advance a similar policy in England, Scotland and Wales.

Forbes eschewed politics, focused on the technical aspects of forestry and kept his head down. This was probably wise because of the violent political storms that raged at this time. The First World War was punctured by the 1916 Rising and the subsequent rise of Sinn Féin and the Irish Republican Army. Afforestation was a low priority during the difficult and violent birth of the independent Irish Free State in the southern twenty-six counties and the partition of six counties into Northern Ireland (which remained in the United Kingdom). Forbes was appointed as director of forestry in the new Department of Agriculture in Dublin in 1922 and remained in office until 1931. He advocated for a continuance of the afforestation policy that had been started before the war. 'The future of woods and plantations in Saorstát Éireann is a matter that ought to concern all sections of the population,' he wrote in the journal of the Department of Agriculture in 1924. 'The time has now come when the Irish Free State must definitely make up its mind whether its woods are to be preserved and extended, or whether it is to become a byword amongst the nations of Europe as the only country without trees.'

The political leaders of the new Irish state heeded his advice – up to a point. They took steps to preserve existing woodlands by enacting new legislation. The 1928 Forestry Act was a watershed. The act made it unlawful to cut down any tree, with certain defined exceptions, without a licence (a provision not enacted in Britain until 1951). Forest owners were also required to replant their land after felling. In theory, the state could now prevent the conversion of woodland to farmland, thus bringing an end to the 6,000-year process that had slowly cleared the Irish landscape. The state also acquired private woodlands through the workings of the Land Commission, which had been set up under the Land Acts to acquire estates and to transfer them to tenant farmers. If the Land

Commission acquired woodlands as part of an estate purchase, it often transferred them to the state Forest Service* for management. Via the Land Commission, the state also continued to acquire bare land for afforestation. The graduates of Avondale (and later forestry schools) did the hard work of fencing off sites, preparing the ground, sourcing and planting seedlings and protecting new plantations from weeds, animals and storms. Between 1922 and 1939 these foresters planted 24,115 hectares of trees.

There were limits to what the government could achieve. Faced by a parsimonious Department of Finance, the budget for afforestation was small. The state's land bank continued to grow, but the rate of planting could not keep pace. Equally, the foresters had their hands tied by a rule about the type of land that could be used for afforestation. Irish farmers were still hungry for land. Politically, it would be suicide for the government to be seen to compete with local farmers for land, or to put good farmland under trees. James Dillon TD highlighted the conflict between forestry and agriculture in a debate in the Dáil on the forestry budget in 1940:

> Now, you have got to choose between trees and men. It may be that some of the land on which men can live would yield a larger return in cash if the men were put out and the trees installed in their place, but even if that land did yield a larger return in cash on that basis, I should still advocate leaving the men on that land and displacing the trees.

As a result, the Forest Service was only allowed to acquire the lowest-value land. In practice, forestry was pushed onto the bogs and wasteland, which limited the species that could be grown, affecting how quickly they would grow and their eventual economic return.

At the same time as new forests were being planted, old woods continued to be cleared. In the 1920s and 1930s the slow

---

* The state bodies responsible for forestry in Ireland had different names at different times but for simplicity I use the term 'Forest Service' throughout.

euthanasia of the Ascendancy class continued, the remaining Anglo-Irish estates came under further economic pressure and estates folded and changed hands. The transfer of ownership was often accompanied by destruction of old demesne woodlands. For example, when the banks foreclosed on Castle Durrow in County Laois in 1922 and Lord Ashbrook took his family off to England, the estate was acquired by a Mr Maher from County Kilkenny, who was attracted by the timber reserves. Over the next six years, teams of men wielding long cross-cut saws worked their way through 263 hectares of oak, beech and ash. (Durrow comes from the Irish *Dearmhagh*, which means 'The Plain of the Oaks'.) When the Second World War broke out in 1939, Ireland stayed neutral but struggled to import goods and had to fall back on indigenous sources of energy and materials, which led to a further burst of felling as old woodlands were raided for timber.

The net result was that there was little change in forest cover in Ireland during the first half of the twentieth century. There were around 120,000 hectares of woodland on the island of Ireland in 1907 and by 1950 this figure had only increased by around 2,000 hectares. Ireland was nowhere near achieving the target of 1 million acres (or 405,000 hectares) of woodland that had been set by the Departmental Committee in 1908. When the UN Food and Agriculture Organization sent a mission in 1950 to report on the state of Irish forestry, it concluded that 'Ireland is the poorest off in forests of any European country'. The biggest change during this period was in ownership. Whereas there had been no public forests in 1907, by 1950 the state (north and south) owned 63,000 hectares, more than half the total. This came about partly through state afforestation and partly through acquisition of older private woodlands, including felled areas that were later restocked.

The real burst of afforestation came in the second half of the twentieth century. It was kickstarted by a man who had one of the most varied careers of any Irish politician of this era – Seán MacBride. A staunch republican, MacBride was the son of Maud Gonne, a legendary leader of the 1916 Easter Rising. MacBride was a prominent member of the illegal Irish Republican Army after independence, serving as Chief of Staff in 1936. After establishing a reputation as a barrister he founded a socially radical political party, Clann na Poblachta, and brought them into a new coalition government in 1948 in which he served as Minister for External Affairs. Later on, after leaving Irish politics, he became secretary-general of the International Commission of Jurists and chairman of Amnesty International, and was awarded both the Nobel and Lenin peace prizes in the 1970s.

MacBride had a fanatical interest in forestry. 'I think I can say that I have been keenly interested in the whole question of the reafforestation of Ireland ever since my childhood days,' he wrote. 'The aim of pursuing an active afforestation policy was an integral part of the Sinn Féin movement in the early portion of the century.' Although as Minister for External Affairs he had no forestry brief, he used his position in the coalition government of 1948–51 to drive a more aggressive policy. He took the fight to the Department of Finance, enlisting the help of the US ambassador to Ireland and tying US financial aid under the Marshall Plan to the forestry programme. The government announced a target of planting 10,000 hectares per year for forty years – a big increase from the 2,425 hectares per year that had been achieved up to that point – with the goal of reaching 405,000 hectares of total forestry in southern Ireland. The government conducted a 'flying survey' to confirm there was enough suitable land to sustain the programme. They invited the UN Food and Agriculture Organization to carry out its study on Irish forestry, which more or less supported the policy targets. The forestry budget was increased and the Forest Service went into expansion mode.

The magical figure of 10,000 hectares of annual afforestation was first reached in 1960. The policy was renewed by successive governments and gradually given a more solid economic underpinning. This was a period of intensive economic planning, led by a more progressive Department of Finance shaped by T.K. Whitaker. The Programme for Economic Development published in 1958 stated that 'forestry could be considered economic if the sale of crops yielded enough to repay the capital cost, with compound interest'. When the department passed their slide rule across the state's past investments in forestry, they found that it would provide an annualised return of no more than 2.5 per cent. This was not very impressive, as the government's cost of borrowing was above this rate. But with cost efficiencies and rising timber prices they expected the investment return to increase to 5.25 per cent in the future, which justified continued investment. The government also took tentative steps to stimulate the development of a timber-processing industry in Ireland. Once started, this became another reason for continued investment in afforestation, as the sawmills and chipboard factories needed a constant supply of timber long into the future to sustain their operations and recoup their investments.

Yet the afforestation programme was still hamstrung by the policy of avoiding competition with agricultural land. The government set a maximum price per acre that the Forest Service could pay to acquire land, which put all but the poorest-quality land out of reach. One official writing in 1963 observed that there was 'a gradual lowering in the average quality of the land being planted' as time went on. If afforestation was to happen, it would have to happen on the blanket bogs, or what was known in farming circles as 'rough mountain grazing'. These were places where trees did not want to grow. Fortunately, new techniques were emerging at this time in Britain for planting peatland. Foresters used heavy ploughs drawn by crawler tractors to improve drainage and to build mounds into which trees could be planted. They applied phosphorus fertiliser to improve soil fertility and provide nutrients for trees to grow. It was a brute force attempt to, finally, turn Irish wastelands into gold. State afforestation from the 1950s onwards took place mostly on these blanket bogs, mainly in the western counties or in upland areas of other counties. Later, in the 1980s, the government encouraged the Forest Service to plant trees on raised bogs in the midlands that had been cut over for fuel by Bord na Móna (the Turf Development Board).

Under the control of a separate administration, Northern Ireland followed a similar path. In 1910 Forbes had bought the 80-hectare Ballykelly wood in County Derry, which became the nucleus for the first state forest in the six counties. From the 1920s to the 1940s the forestry division of the Northern Ireland government was mostly staffed by Scottish foresters, and there were close links with Scottish forestry schools. State afforestation took off in the 1950s, just like in the south. Northern Ireland foresters tried just as hard to establish trees on bogs rather than competing for good agricultural land. Northern Ireland even had its own 'Knockboy' in 1955, when there was a disastrous attempt to establish 60 hectares of forestry on Rathlin Island, one of the most windswept corners in the land. All that is left now is some stunted scrub.

By the early 1980s, state afforestation efforts on both sides of the border had greatly expanded Irish woodlands. The area under forest reached 441,000 hectares, or 5.2 per cent of the total land area. This was state-dominated forestry: 78 per cent of the area was owned by the state. Most were young plantations less than thirty years old. In contrast, most private woodlands were more than fifty years old (although this included a large proportion of low-value scrub). Despite a slow start, the country had finally delivered on the target of achieving 1 million acres of woodland first set by the Departmental Committee in 1908.

THE TYPE OF FORESTRY practised in the twentieth century was different from the gentlemanly forestry of the previous century. Irish foresters embraced 'scientific' forestry. The modern science of forestry had been developed, to a large extent, in Germany in the nineteenth century. Responding to contemporary concerns about timber shortages, and rejecting traditional peasant management of natural woodlands, German foresters sought to establish plantations on an industrial scale. They favoured monocultures of fast-growing conifers such as pine or Norway spruce; they planted saplings grown in nurseries rather than relying on natural regeneration; and they cleared entire plots when they reached maturity, before replanting. German forest academies developed methods to calculate tree growth, harvest volumes and the division of forests into felling units, so as to deliver a predictable and sustainable yield of timber. This was rational forestry based on formulas and tables. Some of these German foresters went to work for the Indian Forest Service and to establish forestry schools in Britain, bringing these new ideas with them. From there, the ideas percolated to Ireland.

Indeed, the government in Dublin went straight to the source in the 1930s by appointing German forester Otto Reinhard as head of the Irish Forest Service. He brought professionalism and strong technical knowledge during his four years in the role. A dapper former German army officer, who was awarded an Iron Cross in the First World War, he was described by an Irish army report as 'very charming' and something of a 'gay dog'. He joined the Nazi party while in Ireland in 1939 but was stranded in Germany while on holidays when the war broke out, which saved the Irish government the headache of dealing with a Nazi forester on its payroll while trying to preserve a precarious neutrality.

Over the course of the twentieth century Irish foresters, north and south, took these Germanic principles of scientific forestry and tailored them to Irish conditions. A standard playbook emerged, and it is largely followed to this day. Agricultural land is cleared and prepared for planting by using various methods of drainage. A single tree species is chosen for each plot. Seedlings are grown in nurseries for a couple of years, uprooted and then transported to the site for planting (when they are known as 'whips'). Trees are planted in high densities to encourage straight growth without excessive branching, as this produces the best timber. Between 2,500 and 5,000 trees are planted in each hectare – an area equivalent to one and half football fields – depending on the species. (So if you hear companies making grand claims about planting thousands of trees, remember this does not mean a large area of woodland.) A fence is erected around the plantation to keep out grazing animals. For the first four or five years, the forester actively maintains the site, clearing vegetation that could strangle the young trees.

After the forest is established, not much happens for a while. So long as the forest is fenced, a forester can leave it alone to grow. But eventually a forest will need to be thinned to maximise its final value. For fast-growing conifers, thinning may start around fifteen or twenty years after planting and it is usually carried out three or four times. The purpose of thinning is to remove poor-quality and small stems and to free up space for those trees that will grow to maturity. Thinning usually requires the construction of roads and tracks so that harvesting machinery can access the site and take timber out.

Then, the big day comes. Like teenagers, trees grow quickly when young, increasing in height, diameter and value. But then the growth rate of the forest slows, as does its growth in value. At a certain point, the forester decides to clear the forest. In the old days, this was done by hand with chainsaws. Now, a harvest machine come in, cuts the trunks at the base, strips the trees of their branches and foliage and deposits sawlogs for collection. The entire area is clear-felled. The land is then replanted – as required by Irish law – and the whole cycle begins again, in a second rotation. This 'clear-fell replant' system is easy to execute and provides predictable volumes of timber. It is industrial forestry, in the sense that the primary goal is to produce material for industry. Or, to use an ugly phrase, it is tree-farming.

Over the course of the twentieth century Irish foresters embraced a new range of exotic species to manage under this system. They came from the north-west corner of North America. The cordilleras, a network of mountain ranges on the west of the American continent, were a rich hunting ground, as the moist, cool climate was similar to Ireland's. Ireland's gentleman foresters had begun experimenting with these species in the 1800s, and they were planted by A.C. Forbes at Avondale in systemised trials. Three of these conifers would become the foundation for modern Irish forestry – Sitka spruce, lodgepole pine and Douglas fir.

Douglas fir (*Pseudotsuga menziesii*) is the most important forest species in western North America, where its timber is known as Oregon pine. Despite its common name it is not a true fir tree but belongs to the pine family, hence the word 'pseudo' in its name. It is a tall evergreen tree that can reach impressive heights and live for 1,000 years. (The tallest tree in Ireland is a Douglas fir growing at Powerscourt in County Wicklow and measuring 61 metres.) The trunks have resin-filled blisters and their leaves release a sweet citrus smell when crushed. Douglas fir is capable of high yields, but it requires a fertile, free-draining soil. Such soils were rarely available for afforestation in Ireland, so the tree was only planted in small numbers. It produces a tough timber that is ideal for garden furniture or boats. For decades, its most valuable use in Ireland was as electricity transmission poles for the Electricity Supply Board.

Lodgepole pine (*Pinus contorta*) was more widely planted, especially in the 1950s. It now accounts for almost 10 per cent of the total forest estate in the Republic of Ireland. Lodgepole pine got its name because nomadic Native Americans used the light, straight wood to support their lodges or tipis. It forms a medium-size, slender evergreen tree. It is an undemanding species that can grow on infertile peat soils on exposed sites. If nothing else would grow, Irish foresters turned to lodgepole pine. But it is problematic. It grows across a wide area in North America and consists of many varieties. Southern coastal varieties, from southern Washington state and northern Oregon, grow quickly but they produce crooked, low-value timber. Other North

America varieties produce straighter stems, but they have slow growth rates. When planted in Ireland, lodgepole pine is also subject to a variety of diseases and insect pests. Because of its low yield and poor quality, most lodgepole pine in Ireland is harvested for pulpwood and used to make panel boards such as medium-density fibreboard (MDF).

Sitka spruce (*Picea sitchensis*) was by far the most important American tree to hitch a ride to Ireland. It now accounts for 51 per cent of all woodland in the Republic of Ireland. Although planted as a specimen tree on estates, its real value became apparent when it was trialled by Forbes at Avondale. Sitka spruce is an adaptable species that can be grown in a wide range of climatic conditions and soil types, but it is particularly suited to Ireland's wet mineral soils, where it can achieve prodigious growth rates. It is also resistant to pests and diseases. Sitka spruce produces straight, flat needles that are sharp to the touch – which is why it does not make a good Christmas tree. Larger trees are used to make planks for construction, while smaller logs are used to make pallets or ground up to make panel boards. In its native range, which extends from California to Alaska along the Pacific coast, Sitka spruce can live for hundreds of years and grow to a towering size – it is the third-largest tree in the world, behind the redwoods and Douglas fir. In Ireland, it has more prosaic associations. It is grown in dense plantations and typically felled after thirty or forty years. It is the Peter Pan of Irish forestry, perpetually kept in a fast-growing juvenile state and rarely allowed to grow to maturity.

Sitka spruce is now the workhorse of Irish forestry. It has the fastest growth rates, is the species around which sawmill investments have been made and it delivers the highest financial return to the landowner. If some foresters had their way, they would plant Sitka spruce from fence to fence and little else. In the early 1990s, 80 per cent of the newly afforested area was planted with Sitka spruce; by 2019 this figure was still almost 70 per cent. There are only two situations where Sitka spruce is not the conifer of choice. One is land with deep peat soils, where the tree grows poorly and can be blown over by wind. The other is low-lying areas, especially in the midlands, that suffer from late spring frosts, as these frosts can damage its buds – the hardier Norway spruce, a native of Scandinavia, is planted there. Sitka spruce has found similar favour in Britain, where it was planted widely in the twentieth century. Indeed, there are now more Sitka spruce trees in Britain and Ireland than there are in its natural range in North America.

STATE AFFORESTATION BASED ON these 'scientific' methods was sustained in the Republic through the 1960s and 1970s, when the area planted averaged 8,795 hectares per year. But in the early 1980s the effort ran into headwinds. The accession of Ireland and the UK into the European Common Market in 1973 led to a flow of agricultural subsidies to Irish farmers, which raised land prices and made it harder to acquire grazing land for planting. The costs of the Forest Service ballooned in the 1970s because of inflation and the large areas that now required management. Originally, it had been estimated that revenues from timber sales would be sufficient to cover all forestry operating and capital costs by 1982–3. The reality was very different – the Irish Forest Service had a massive deficit each year. There were also growing concerns about whether large areas of state forestry would ever produce an economic return because of the poor growth of trees on difficult sites. The 1980s was a decade of austerity in Ireland and the Exchequer took an axe to the forestry budget. State planting rates dropped. Instead, policymakers looked to the private landowner – and especially the Irish farmer – as the means by which afforestation could continue.

In theory, the promotion of afforestation by private landowners had been part of official policy from the beginning. The governments in Dublin and Belfast offered small grants to landowners for this purpose, but they had little impact. In the Republic the amount of private land planted between 1922 and

1955 averaged less than 100 hectares each year. The figure was even lower in Northern Ireland. The total amount of woodland in private ownership on the island more than halved during this period, as old estate woods were cleared and surviving woods were transferred to state ownership by the Land Commission.

A few descendants of the old Anglo-Irish landlord class clung on. Writing in 1957, T. McEvoy described how they had 'survived depressions, land wars, the implementation of the Land Acts, the workings of the Land Commission and the imposition of inheritance taxes'. They were characterised by 'tenacity, an enduring commitment to the land and, in some cases, considerable frugality'. Trees were important to them as 'both symbols of good stewardship and as links with the past', representing 'a patrimony and a tradition to be handed to one's heirs'. When managing their forests, they combined economic objectives with 'aesthetic, amenity and wildlife considerations'. Yet these landowners were sidelined by the state, received little encouragement to expand and were usually ignored when they offered advice that conflicted with the official policy favouring industrial conifer plantations.

The new era of private planting from the 1980s focused on a different constituency – the small Irish farmer who had been the beneficiary of land reform. The initial stimulus and funding came from policymakers in Brussels, who flipped the old calculus that had governed competition between forestry and agriculture. Before, Irish politicians were afraid to divert land towards forestry as they wanted to help farmers produce as much food as possible on every available acre. Now the European Economic Community (EEC) was groaning under beef and butter mountains, milk lakes and grain piles. Reducing agricultural production, not expanding it, was the imperative. At the same time, the EEC imported 60 per cent of its timber needs. Putting more farmland under forestry, therefore, would achieve the triple goals of reducing agricultural production, increasing timber supplies and providing new livelihoods for rural communities.

In 1981 the EEC-funded 'Western Package' was made available to the poorest counties in the west of Ireland to encourage afforestation. The scheme was later expanded to the whole of the Republic and continued through a series of national Forestry Programmes, each more generous than the last. By 1989 the essential features of the grant scheme were defined, and they have remained in place ever since. The government offered grants to private landowners to cover the costs of establishing a new woodland. Even better, the government paid an annual premium to the landowner for the first fifteen to twenty years to compensate for lost agricultural income, all tax-free. For many years farmers qualified for a higher premium than non-farmers, but in 2014 this distinction was scrapped and all recipients were paid the higher amount. Additional grants were later provided to help cover the cost of building new roads to access plantations and to support the establishment of native woodlands. This was the most generous grant scheme for private afforestation anywhere in Europe. It was made possible by continued funding from Brussels, although the Irish Exchequer took on a greater share as time went by.

The new grant scheme had an immediate impact. The annual rate of private afforestation rose from 275 hectares in 1981 to 3,215 hectares in 1987 and peaked at a mighty 17,343 hectares in 1995. Irish farmers were responsible for most of this planting. They took their worst land, either rough grazing land on hillsides, or the wetter fields that posed problems for raising livestock, and signed up for the guaranteed grant income. They planted small plots averaging less than 10 hectares. For a while, investors and pension fund managers such as Allied Irish Bank and Irish Life got in on the act too, acquiring bare land for planting or buying newly established plantations from farmers. They were responsible for half the forestry planted in 1986, although their share later dwindled as farmers embraced the scheme. An ecosystem of private forest companies developed to carry out planting and to manage the young forests. On a smaller scale, in the 1990s there was an increase in private afforestation in Northern Ireland too, encouraged by UK grants and tax incentives.

In the Republic, the state foresters also took one step towards the private sector around this time. It was an era of privatisation and scepticism towards state-led economic development. In 1989 the Irish government created a new semi-state company called Coillte Teoranta to take over all existing state forests. It was supposed to operate on commercial lines, freed from bureaucracy and political interference, although the government was the ultimate shareholder. Coillte took over 396,000 hectares of state forestry, which it continues to manage today. Most Forest Service staff were transferred to the company. In the drive for commercialisation, Coillte developed a new panel-board factory near Waterford to turn low-grade pulpwood into valuable final products and set up more professional systems to sell logs from its forests to private sawmills.

Initially, Coillte continued to plant new forests. It was able to avail of the same afforestation grants offered by the government to private landowners. The rate of state – or, more correctly, semi-state – planting actually increased from a low of 4,625 hectares in 1985 to 7,855 hectares in 1995. But in 1999 this activity came to a shuddering halt. The European Court of Justice ruled that Coillte, as a state-owned company, could not receive grants from official sources because of restrictions over state aid. Deprived of the grants that it needed to cover its costs, Coillte stopped establishing new forests and focused on managing its existing assets instead, only carrying out replanting after areas were clear-felled, as required by law. Henceforth, when it came to woodland creation, the private sector was the only game in town.

The shift to private afforestation continued the momentum that had been started in the 1950s. Between 1950 and 1985, foresters in the Republic planted 272,000 hectares of mostly state woodlands; over the next twenty years, they planted 286,000 hectares of mostly private woodlands. The methods were much the same. Private afforestation was primarily based on conifers, mostly Sitka spruce, and followed the same template of 'scientific' forestry laid down by the state foresters earlier on. As a result, today more than two-thirds of Irish forests are conifers, and just under one-third are broadleaves. Because of the nature of this afforestation drive, Ireland's forests look very different from the rest of Europe. Ireland has the second highest proportion of forestry in plantations (behind Malta), the largest proportion of non-native tree species (69 per cent) and the least amount of forests with protected status. But there was a rebalancing of forest ownership that brought Ireland more in line with European norms. In the early 1980s, 78 per cent of Ireland's forests, north and south, were state-owned. Today, around half the forests are owned by the state and the rest are private. There are now more than 50,000 private owners scattered throughout the island, all with a stake in forestry. Once a nation of farmer proprietors, Ireland now has a sizeable constituency of forester proprietors.

BY 2020 THE TOTAL AMOUNT of woodland in Ireland, north and south, had grown to 898,000 hectares. Forests now occupy 10.7 per cent of the total land area of Ireland, north and south. This is well below the European Union average of 45 per cent, but it is more than Malta, neck and neck with the Netherlands and not far below neighbours such as Britain and Denmark (which have around 15 per cent forest cover). Ireland is no longer languishing on its own at the foot of the forestry table.

The reforesting of Ireland over the last 100 or so years is a singular achievement in many ways. From having almost no woodlands in 1920, Ireland now has a real forestry sector that produces large volumes of harvested wood each year, supporting thousands of jobs. The country is a net exporter of wood products again (in volume terms). Forest cover is now at the highest level for at least 400 years. A new forest ecosystem is emerging, and old species are returning. In 2007 birdwatchers were thrilled to discover that woodpeckers had appeared in Ireland after being extinct for more than two centuries. You can hear their tap-tap-tap in Wicklow woodlands and even among the trees of Cabinteely Park, down the road from my childhood home. The Irish landscape has been transformed. A.C. Forbes, Seán MacBride and the other pioneers of Irish forestry would be amazed – and a little proud.

GRINDING TO A HALT

AT THE TURN OF the millennium, the future of Irish forestry looked bright. The trees that had been planted on private land were, for the most part, growing better than expected. Wet marginal farmland – away from peat bogs – was perfect for growing conifers, especially Sitka spruce. Ireland benefited from a mild, humid climate, a long growing season and an island's natural defences against arboreal pests and diseases. Average growth rates for spruce in Ireland were one-fifth higher than in Britain and two to three times higher than in Finland or Sweden. It turned out that Ireland was one of the best places in Europe, if not the world, to grow trees.

Study after study showed that the economic returns from planting trees were much better than trying to raise cows or sheep on marginal grazing land. Irish farmers were struggling as commodity prices failed to keep up with the rising costs of production. Many were forced to take off-farm jobs, turning farming into more of a hobby. Government advisers lined up to persuade them to 'go into the forestry'. It was the economically rational thing to do. Other studies identified plenty of land to plant on. Niall Farrelly of Teagasc, the state agriculture research authority, calculated there were 1.3 million hectares of marginal farmland in the Republic that could be planted to forestry without greatly compromising agricultural productivity. This was almost twice the size of the existing forest estate.

There was plenty of demand for the timber from these forests. After a rocky start, and some bankruptcies in the 1980s, the Irish timber-processing industry had developed into a force that could compete on global markets. Sawmills bought larger-diameter logs, known as sawlogs, and turned them into planks and boards for use in construction. This was by far the highest and best use for timber, commanding the best prices. Medium-diameter logs were turned into pallets or fencing. Three panel-board factories in counties Leitrim, Tipperary and Kilkenny purchased smaller logs and branches, turned them into pulpwood and manufactured high-tech panel boards such as medium-density fibreboard (MDF), oriented strand board (OSB) and doors. Approximately €200 million was invested into modernising and expanding these sawmills and factories in the early 2000s. During Ireland's manic property boom in the early 2000s most of these timber products were used at home, but, following the economic crash around 2008–9, the processing industry showed its adaptability by turning on a penny and exporting a large proportion to the UK and other markets. The value of Irish timber exports reached €450 million in 2018.

There is a popular misconception that wood from Irish conifers is of poor quality because 'the trees grow too fast'. This is not the case. This type of softwood is not going to be used to make fine furniture or laid down as flooring in a trendy loft-style apartment, but it has plenty of uses in construction and manufactured products. Throughout this period, forest owners were pleasantly surprised by the prices they could get for their timber.

There was also a new factor that supported continued afforestation – climate change. For a century, the rationale for planting trees had been job creation and economic development. Now, forestry had an extra value because of its ability to store carbon. This was not just an abstract environmental issue; cold, hard cash was at stake. Under EU Directives, Ireland would be fined if it did not keep its greenhouse gas emissions below certain levels by 2020 and 2030. The problem was that the shift to renewable energy was behind schedule, the economy was still powered by fossil fuels, and the growing number of dairy cows were belching more and more methane into the air, creating a situation where Ireland had the third highest per capita emissions of any EU member state. Forestry was a saviour, as it stored carbon that could offset these emissions. This could be worth hundreds of millions of euros per year to the Irish government, as otherwise it would have to pay fines to Brussels or purchase credits from other countries. It meant the state might receive a payback from its historic investment in forestry in a way that Seán MacBride or other advocates could never have expected.

Climate change policy also created new demands for timber products in the form of bioenergy. The EU Directives on climate required Ireland to secure a large share of its electricity and heat from biomass, and the biggest potential source of biomass was wood from forests. Bord na Móna started to wind down its peat mining operations in the midlands – which spewed out even more carbon emissions than coal – and to convert its power stations to burn woodchips and pellets instead. Factories, hospitals and buildings started to install wood boilers to provide heat. Suddenly, the small-diameter logs that had gone to pulpwood had a new market. Pulpwood prices tripled, which meant that thinning operations in young forests were more profitable. Government projections for timber markets showed demand outstripping supply for many years, largely because of the impact of bioenergy.

As a result, the Irish government doubled down on the afforestation policy. In 1996 the government published a strategic plan for the development of the forestry sector, *Growing for the Future*. It set a national planting target of 25,000 hectares per year until 2000 and 20,000 hectares per year from 2001 to 2030. The goal was to reach 17 per cent forest cover, or 1.2 million hectares, in the Republic by 2030. Most forestry experts believed that this would be easily achieved by private landowners encouraged by government grants.

Yet no sooner had these ambitious targets been set, actual planting rates started to fall. In 2003, for the first time since the 1980s the annual amount of afforestation fell below 10,000 hectares. The figure dropped to 6,653 hectares in 2011 and it has been falling ever since. The targets under subsequent

Forestry Programmes were repeatedly ratcheted down to reflect these new realities, but in every period the planting area fell well short of the targets, even though grants increased. By 2020 the total area planted to new forests had fallen to a woeful 2,434 hectares. In 2021 it dropped to 2,016 hectares. What had gone wrong? Why was the great Irish afforestation effort grinding to a halt?

ONE REASON FOR ANAEMIC planting rates was the attitude of Irish farmers. It turned out that they were less enthusiastic about forestry then they should be. Fathoming the mind of the Irish farmer and understanding why he (and it was usually a 'he') was so reluctant to plant trees became a sort of cottage industry to academics, who went to work with surveys, interviews, anthropological case studies and complex econometric models. The Irish farmer became a much-studied species.

One important factor in explaining farmers' behaviour was the impact of warring subsidies. While the government was offering grants for forestry, increases in agricultural subsidies in the late 1990s – through the Rural Environment Protection Scheme and an extensification premium paid to farms with low stocking densities – created incentives for farmers to keep marginal land in livestock production. Farmers also feared that by putting land under forestry, they would lose out on future area-based subsidies under the European Common Agricultural Policy. Indeed, the quirks of the subsidy system may partly explain the spike in private planting in the 1990s. At that time there were fixed quotas for milk production and dairy farmers could only expand production if they acquired quotas from other farmers. A simple way to do this was to buy poor-quality land from a retiring farmer, obtain his milk quotas, then take the land out of production and plant trees. When quotas were abolished, this rationale went away too. Irish farmers have become experts at 'farming the grants', adjusting decisions on land use as the rules change.

After a few years most of these clashing incentives were ironed out and agricultural subsidies modified to ensure that a farmer embarking on forestry was not penalised. Yet planting rates continued to fall, even as economists swore to the superior returns from forestry on many of Ireland's beef and sheep farms. This points to deeper reasons behind the reluctance to plant. And they are cultural and social, rather than economic.

The reality is that most farmers are hostile to forestry because 'it simply isn't farming'. One survey found that 84 per cent of farmers would never put land under forestry no matter how generous the financial incentives from the state. The vast majority of farmers will only consider planting land that is 'good for nothing else'. To many farmers, afforestation is associated with 'loss': loss of open agrarian space, loss of traditional farming work and, ultimately, loss of rural identity. Someone who turns to forestry is seen as a failed farmer. The reluctance to plant is especially strong among farmers with small holdings, who feel that all their land is needed for agriculture, and among farmers with children, who want to pass on as large a farming unit as possible. The 'land hunger' engrained in Irish farming communities for hundreds of years has not gone away. The permanent nature of the change to forestry – by law, once planted, forest land cannot be converted back to farmland – also holds farmers back. This is a decision for their lifetimes, and for those of future generations.

There was also broader rural community opposition to forestry, at least to the type of spruce-dominated monocultures that had sprung up around the country. As we have seen, there was a long-standing antipathy towards forestry in Ireland, stemming from the time that forestry was associated with Anglo-Irish landlords. Opposition to commercial forestry began to coagulate in the 1980s. Critics lamented the march of conifers that blotted out the traditional open landscape of rural Ireland. Afforestation was associated with the abandonment of farms, the depopulation of rural areas and the breakdown of local communities. Trees were replacing people – and the people who were left behind didn't like it.

This opposition was most visible in Leitrim. The county, with its drumlin hills and heavy, wet soils, has some of the best conditions for growing Sitka spruce. By 2017 almost 19 per cent of the county was covered in forestry (compared to an average of 11 per cent across the Republic of Ireland). But this wave of afforestation provoked hostility from the start. In the 1980s the local branch of the Irish Farmers Association came out against tree-planting. On one occasion, angry locals burnt forest machinery. In the 2010s a 'Save Leitrim' campaign was formed to protest against the creeping tide of conifers. On 30 January 2019 the group organised a protest in Dublin billed as 'Communities not Conifers', bussing 100 supporters to march outside Dáil Éireann. Leitrim even made it into *The Economist* magazine as an example of a worldwide trend of local opposition to afforestation.

One of the founders of Save Leitrim is Edwina Guckian. A musician and dancer, she is at the fore in reviving the Irish tradition of sean-nós dancing, which she learned from her grandparents. When, as a child in Leitrim, she heard that a forest would be planted beyond her family's farmhouse, she imagined a sylvan paradise, full of life, the world of Bambi. Instead, she got serried ranks of Sitka spruce towering over their home. Her family's cattle would go inside, get lost and not come out for weeks, starving. 'It's like a wall around you, dead, darkness,' she told a reporter. 'It's suffocating. We're losing the landscape.' Another member of Save Leitrim, a local farmer called Jim McCaffrey, complained how 'the forest closed in bit by bit'. It was 'a death sentence for the townlands'. 'The sun rises late and goes down early,' he continued. 'And when they harvest the trees it is like Armageddon.'

The Save Leitrim campaigners made an unlikely alliance with an increasingly important group in the debate about forestry – environmentalists. From the 1980s onwards, forests were no longer viewed solely through an economic or social lens, but were also judged on their environmental impacts. And they were found wanting. Ecologists attacked conifer plantations on many grounds: for acidifying waterways, for suppressing biodiversity, for degrading soils, for endangering precious habitats, for ruining landscape aesthetics. They complained about 'the dark tide of conifers enveloping the uplands'. A thicket of non-profit organisations sprung up to condemn spruce plantations and to champion native woodlands as an alternative. It began with Crann, founded in 1986 by an Australian woman, Jan Alexander, from her base in Leitrim, with the aim of 'releafing Ireland'. Other organisations have sprouted since: the Tree Council of Ireland, the Native Woodland Trust and the Woodland League, to name a few.

After some years in the wilderness, Irish environmentalists began to receive powerful support from the policies, directives and judicial decisions of the European Union. Brussels had provided the funding to kickstart private afforestation in Ireland in 1980s. But, in a case of buyer's remorse, European policymakers now started to place constraints on the plantation-based model of forestry championed in Ireland.

The first intervention was the 1992 EU Habitats Directive, which required Ireland to conserve areas of high biodiversity value and to protect endangered species. The Irish government designated 430 conservation areas around the country as Natura 2000 sites. One species – the freshwater pearl mussel – received special attention. Capable of living up to 140 years, and therefore Ireland's oldest animal, this mussel can only survive in pristine river systems. It is sensitive to any increase in silt or nutrients in a river, the sort of thing that can be caused by afforestation or forest-harvesting operations. Whole river systems – such as the Blackwater in County Cork – were designated as protected areas. Before issuing forestry licences, the inspectors of the Forest Service now had to assess whether the proposed activities would have any impact on Natura 2000 sites, either directly or indirectly, via streams that flowed into these sites. As the Natura 2000 sites were scattered across the island, this affected a large area.

In 2009 a new EU Birds Directive came into effect. This focused attention on another endangered species, the hen harrier. A raptor that lives on heather moorland, it was threatened by the loss of habitat to forestry. By 2015 fewer than 150 breeding pairs were left in the Republic. While they can use young conifer plantations for nesting and foraging, as plantations mature and the forest canopy closes, this habitat becomes unsuitable. Six Hen Harrier Special Protection Areas (SPAs) were designated in the Republic in 2007, covering a total of 167,000 hectares. All new forestry was banned from these areas as well.

These were not the only areas that were placed off limits. The combined influence of Irish environmentalists and EU Directives has brought a complete turnaround in perceptions of Irish bogs. Long regarded as 'waste land', and the subject of afforestation fantasies for more than a hundred years, ecologists now regard them as precious habitats that need to be preserved. This is partly due to a broadening of horizons. Ireland has a lot of bog – too much for many Irish people – but in European and global terms it is rare, and it supports unique and endangered species (such as the hen harrier) that are not found elsewhere. Its preservation is therefore a European priority, and this has to be translated into Irish law. Irish bogs have also taken on new importance because of climate change. They are huge stores of carbon, and the drainage and preparation of land for forestry, by drying the soil and bringing oxygen into contact with the buried organic matter, releases this carbon. Carbon accounting for forests on peatland soils is complicated, as the growth of trees above ground can compensate for the soil carbon losses below, but the consensus is that it is best not to disturb intact peatland. Since 2011 the Irish Forest Service has prohibited new afforestation projects on peat soils deeper than 50 cm and limited the amount of 'unenclosed land', in other words rough grazing, to less than one-fifth of any new project. Most of the areas that were planted with state forestry from the 1950s to the 1970s, and many that were planted with private forests in the 1980s and 1990s, would not receive permits now.

This does not solve the problem of what to do with legacy forests on peat soils. A lot of these forests are of poor quality and low economic value. In some cases, Sitka spruce or lodgepole pine did not grow properly because of a lack of nutrients or waterlogging – they are permanently 'in check'. If the trees did grow, they are at huge risk of being blown over by storms, as their shallow roots do not have a strong grip in the wet peat soils. Lodgepole pine is a particular problem and has lived up to its Latin name, *Pinus contorta*. It is prone to crooked stems and something called 'basal sweep' – where the wind blows over a sapling, which then grows upwards in a curved shape – producing wonky logs that have little value to a sawmill.

Many of these peatland forests cannot be harvested and are in limbo. Some are on such environmentally sensitive sites that clear-felling would cause too much damage. Some cannot be accessed because it would be too destructive to bring in heavy machinery over the wet, fragile soils – these forests are cut off, as if on an island. Other areas of peatland forests suffer from 'negative stumpage': this is where the costs of harvesting and extraction are greater than the revenue that can be expected from selling the timber. The economically rational choice is not to harvest at all. Finally, there are areas that have positive stumpage but where the small profit from harvesting would be more than outweighed by the costs of replanting forests for the next rotation: the Forest Service is reluctant to waive this replanting obligation because any area that is felled and not replanted is counted as deforestation, which goes in the debit column of the national carbon accounts. When Dermot Tiernan of Coillte examined this subject in 2007, he was candid about the scale of the problem. He concluded that only 11 per cent of western peatland forests could deliver wood production that was economically and environmentally justifiable.

Across Ireland, north and south, 346,000 hectares of forest are found on deep peat. If Tiernan's assessment holds true across all these peatland forests, it means there are more than 300,000 hectares that cannot be harvested for economic or environmental reasons, as things stand. To put it bluntly, the dirty secret of Irish forestry is that around one-third of the forests are no good. The island is full of zombie forests that shouldn't be there in the first place but can't be cut down because it would cost too much or damage the environment. No one is quite sure what to do with them. This is not quite a fiasco on the scale of Knockboy, but it is an ecological and economic disaster nonetheless. Ultimately, we can now see that Irish foresters lost their battle with the bogs. The dream of turning brown bog into green gold has turned into a nightmare.

ENVIRONMENTAL CONCERNS have greatly limited the places where new commercial forests can be planted. But environmentalists, and local community groups such as Save Leitrim, are not against all trees per se. They cherish the pockets of native woodland that exist around the country and champion the planting of broadleaf species rather than fast-growing conifers such as Sitka spruce. These more natural forests support a greater range of biodiversity, improve the fertility of soils and blend more aesthetically into the landscape. Opinion polls show that the public is also in favour of native woodland. We enjoy the seasonal cycles of leaf and colour and the greater openness found under the canopy. The true Celt, it seems, feels a resonance with the sacred trees of the past. Is this the answer to Ireland's forestry problems? Why not plant more native woodland? Why not reorient the forest sector away from conifers and towards broadleaves?

Broadleaves occupy 31 per cent of the forest area in Ireland, north and south, or about 3 per cent of the whole land area. Not all broadleaves are native – there are thousands of hectares of non-native beech and sycamore, for example. And not all natives are broadleaves – let's not forget yew, juniper and Scots pine. Also, when people hear the phrase 'native woodland' they often think of 'ancient woodland'. There is something romantic about the notion that there are sites that have always been under a cloak of trees, in an unbroken link to the primeval wild woods that greeted the first human settlers in Ireland. It makes us feel that, by stepping into these woods, we can somehow step into the past.

The reality is that there are few ancient woods and no wild woods left in Ireland. Between 2003 and 2008 scientists led by Philip Perrin and John Cross carried out a National Survey of Native Woodlands in the Republic on behalf of the Irish National Parks and Wildlife Service. It was one of the largest ecological surveys to be completed in Ireland, involving the study of 1,320 separate woodlands. Using a form of digital tracing paper, the researchers laid down all evidence of historic woodlands, starting with satellite imagery of contemporary forests, then layering on the first detailed Ordnance Survey maps from 1830, and then adding every surviving map before that, going all the way back to the Civil Survey of the 1650s, to see how far back they could trace forest cover. They identified just 16,674 hectares of land, or 9 per cent of the total broadleaf area, that was under woods since 1830. And they could only find evidence for 6,021 hectares of woodland that stretched all the way back to the 1650s. When a similar exercise was carried out in Northern Ireland, the team found only 550 hectares of woodland that could be traced back to the 1600s. The amount of woodland that can claim an unbroken succession with the Gaelic past is vanishingly small.

One reason is the state forestry policy carried out in the twentieth century. As the old Anglo-Irish estates were broken up and transferred to the Land Commission, demesne woodlands were passed to the Forest Services in the Republic and Northern Ireland. The state foresters, more used to working on unforgiving peat bogs and hillsides, eagerly accepted these sites as they generally contained better soils where trees would grow much faster. Rather than preserving the semi-natural woodlands that came with the transfers, their first instinct was to clear the broadleaves and establish plantations. By the 1960s the Forest Service in the Republic had cleared 29,000 hectares of formerly private woods and planted non-native conifers in their place. In Northern Ireland around one-third of ancient woods that survived to the twentieth century were replanted with conifers. It is ironic that government policy goes to such lengths to preserve and expand native woodlands now, when previous policy did so much to hasten their destruction.

Despite this, pockets of ancient and long-established woodlands can be found scattered throughout Ireland. Well-known examples include Glengariff Woods in County Cork, Powerscourt in County Wicklow and Brackloon in County Mayo. After a thorough study, the English ecologist Oliver Rackham concluded that the best-preserved ancient woodland in Ireland is on a peninsula in Lough Ree, County Roscommon. The mixed woodland of St John's Wood includes oaks, hazel, whitebeam and a large number of crab apple trees. Its survival was partly due to the protection of the Knights Hospitallers, crusaders of the order of St John of Jerusalem, whose large thirteenth-century castle stands nearby.

St John's Wood, although ancient, should not be confused with a wild wood. It was subject to intensive management for hundreds of years, as can be seen in the number of coppiced oaks that have regrown. Every woodland in Ireland shows sign of past human intervention. There is no such thing as an Irish wild wood untouched by man. This is true even for the largest area of native woodland in Ireland – the woods that carpet the mountain valleys and lake shores of Killarney National Park in County Kerry.

The Killarney woodlands are the most famous in Ireland. Set in the rugged beauty of the Kerry mountains, and home to the country's largest herd of native red deer, they have been attracting tourists since Queen Victoria paid a visit in 1861. Pollen analysis dating back 5,000 years shows that the Killarney area was originally covered by dense forests of pine, oak and birch. Around 2,000 years ago there was significant human disruption. Scots pine was cut down and eradicated, fires were started (as evidenced by charcoal remains) and the canopy was opened up to graze livestock. During the Williamite wars of the seventeenth century, the lands were confiscated from the second Viscount Kenmare and, according to a later court case, the woods were left 'lying waste and open to every plunderer' in the 1690s. Huge quantities of timber were felled to make staves for casks, charcoal for ironworks and bark for tanneries. In the eighteenth century non-native species were widely planted, including beech and European larch. In the early 1800s, during a spike in demand for oak timber caused by the Napoleonic Wars, there was large-scale felling and coppicing of standing oaks and then extensive replanting that favoured oaks over other species. For example, 49,000 oaks were planted in 1805 in Tomies Wood. In a way, oak was the Sitka spruce of its day.

Today Killarney has the largest area of native oak woodland in Ireland, covering 1,200 hectares. But all is not well. Many of the trees are around the same age, having been planted or regenerated in the early 1800s. The old trees are dying, and there is little natural regeneration because of overgrazing by deer, cattle and sheep and an invasion of rhododendron bushes in the understorey. In an important study, Professor Fraser J.G. Mitchell of Trinity College Dublin concluded that 'these woodlands are a feature of the cultural, rather than the natural, landscape'. They are how they are because they were managed in a particular way two hundred years ago. Moreover, their current structure and composition is 'not sustainable without significant intervention and manipulation'. Rather than being relics of primeval forest, the Killarney woods have been shaped by the felling and planting decisions of previous generations, and their evolution will be shaped by the decisions of current and future generations.

The history of Killarney shows how it is impossible to disentangle human influence from Ireland's native woodlands. Indeed, most native woodlands in existence today were planted by human hand rather than springing spontaneously from the earth. And most were planted very recently: 61 per cent of the broadleaves in the Republic are less than 30 years old. Since the early 1990s a combination of stick and carrot from the Irish government has led to an increase in the planting of broadleaves. Responding to environmental pressure, the Forest Service introduced a requirement that landowners had to plant at least one-tenth of a new forest with broadleaves in order to receive grants to plant conifers. The government also increased the value of grants available for broadleaves so that landowners would receive a higher premium for the first fifteen or twenty years. The proportion of afforestation land planted with broadleaves in the Republic rose from a few per cent in 1990 to a high of 37 per cent in 2011, and has averaged 23 per cent over the last thirty years. It is now common to see strips and squares of broadleaves around Sitka spruce plantations, especially close to roads or streams. The 'dark tide of conifers' is interrupted by patches of deciduous colour, although the compartmentalisation can look just as unnatural.

Yet broadleaves are unpopular with the forest industry and those who seek an economic return from forestry. They cannot compete with Sitka spruce on any standard financial metrics. When the National Survey of Native Woodlands in Ireland was carried out, researchers found that the woodlands had very little timber of value. Less than 3 per cent of the trees studied were of merchantable size and quality. Growing high-quality hardwoods is much harder than softwoods, as the trees are susceptible to defects such as forking, heavy branching and bending. Broadleaves require active management in the form of thinning and pruning over decades to produce the long, straight, unblemished logs that industry needs. This management is costly, few people do it and most broadleaf forests in Ireland show signs of neglect.

The real problem with broadleaves, however, is not quality but *time*. They grow much more slowly than commercial conifers. Whereas a Sitka spruce may grow by 20 to 30 cubic metres per hectare each year and be ready for harvest after thirty years, an oak will only grow by 6 to 8 cubic metres per hectare annually and will not be harvestable for 80–120 years. You might plant Sitka spruce for your pension, but you plant native woodland for your grandchildren and their children. Along the way, the broadleaf woodland will need active management in the form of thinning and pruning. This may produce some revenue through sale of firewood but probably not enough to cover the management costs.

Foresters use discount rates – a sort of interest charge – to convert future costs and revenues into a net present value. Money earned in eighty years is worth less than money in thirty years, for example. Looked at this way, broadleaf and conifer forests have vastly different values. This is why a twenty-year-old, well-stocked, productive hectare of Sitka spruce might be worth €15,000 in Ireland today, but a similar hectare of oak, even if well managed, is probably worth less than €2,000 in purely financial terms. The same harsh economic logic applies at the planting stage. Even factoring in the grants and premiums paid by the government over the first fifteen years, the net present value of oaks or other broadleaves at the time of establishment is much less than what the land would be worth if sold for agricultural use. Anyone who plants broadleaves is doing it for the love, not the money. Or because the government forces them to do it in order to access the real prize – the ability to plant fast-growing conifers.

We all love broadleaves and native woodland, but it is not clear how to make them pay. Moreover, one of the ecological arguments for planting native tree species has taken a hammering over the last decade. Native species are supposed to be better adapted to Irish conditions and more resistant to pests and diseases. But an arboreal pandemic is slowly creeping across the Irish landscape and throttling one of the most widespread and iconic native trees – the ash.

Ash was one of the most widely planted native species during the private afforestation boom of the 1990s and early 2000s. It is faster-growing than oak, and there is a prospect of some decent financial returns after fifteen or twenty years, when stems with the right fluted shape can be harvested and used to make hurleys. Hurling is one of Ireland's most popular traditional sports – and hurleys break a lot in play – so there is always a high demand: around 350,000 hurleys are manufactured in Ireland each year. Only the bottom 1.3 metres of the trunk is used, but the price is ten times what would be earned from selling the timber as firewood.

Ash occurs in almost every native woodland and is the most common large hedgerow tree in Ireland. If humans were to disappear, most of the country would be covered by oak-ash or hazel-ash forests, with ash dominating to a greater extent than in continental Europe, where beech often takes its place. Altogether, ash accounts for 25,280 hectares or 3.8 per cent of the total forest estate in the Republic of Ireland. Woodlands dominated by ash have a rich shrub and herb flora as a result of their relatively open canopy and support a wider diversity of plants than any other woodland type.

In the 1990s a mysterious disease began to afflict ash trees in Poland and Lithuania. Scientists traced it to a fungal disease caused by *Hymenoscyphus fraxineus* that originated in East Asia. With a sense of dread, Irish foresters watched as it slowly spread west across Europe, shrivelling up leaves, wounding bark and killing most of the trees it touched. Ash dieback was first detected in Ireland in October 2012 in a young plantation in County Leitrim that had been stocked with imported trees. Initially, government policy was to fight the disease, paying landowners millions of euros to destroy ash trees in and around affected areas. But, like King Cnut before the waves, and similar failed attempts at eradication in Britain and continental Europe, this action could not hold back the inevitable. Ash dieback (or Chalara, as it is sometimes called) continued to spread, and can now be found in every county. In 2018 official policy changed from fighting the disease to learning to live with it, which is another way of saying let ash trees die.

Ash dieback could dramatically change the Irish landscape in the same way that Dutch elm disease almost wiped out majestic elms forty or fifty years ago. Experts fear that 95 per cent of ash trees in Ireland could succumb to the disease. Grant support for the establishment of new ash woodland was stopped in 2012, which is one reason for the contraction in the area of new broadleaves in following years. The arrival of ash dieback also created understandable nervousness among landowners about the reliability of forestry in general. Forestry had been presented as a long-term but predictable investment. If the native ash, feted in Irish legend as the *axis mundi* around which the world turns, could be wiped out by an invading pathogen, what about other tree species? What would happen if a similar disease struck Sitka spruce, the workhorse of commercial Irish forestry? The economic impact would be enormous. It raised new concerns about the resilience of Irish forests, especially in the face of a changing climate. In the meantime, Irish hurley-makers turned back to imported timber to meet their needs. The 'clash of the ash' in Croke Park or Semple Stadium is often the sound of wood grown in Latvia or England.

ALL THE PROBLEMS ASSAILING Irish forestry seemed to come to a head in 2020 and 2021. The amount of new afforestation dropped to 2,434 hectares in the Republic in 2020 – the lowest rate for seventy years and well below the government target of 8,000 hectares. Landowners who might want to plant were hemmed in by hen harriers, freshwater mussels and the strictures of the EU Habitats Directive, newly enshrined in Irish law. Farmers, never enthusiastic about giving up agricultural land, had lost what little confidence they had in forestry. Moreover, environmentalists and local campaigners had stumbled upon a wrench that they could jam in the machinery of the Forest Service. In 2017 new Irish forestry legislation, to bring the country in line with EU rulings, created a statutory appeals system, allowing any member of the public to lodge an appeal against forestry licence decisions. A

Forestry Appeals Committee with an independent chair now had the final say. There was no fee required for an appeal, and no need to demonstrate a connection with the land in question. More than 700 appeals were lodged between 2018 and 2020. One diligent individual lodged 427 appeals on his own, another environmentalist was responsible for 205 cases, while the Save Leitrim group showed comparative restraint by lodging 59 appeals. Cases were often referred back to the Forest Service for an environmental assessment, but the department had only one ecologist on staff. By Christmas 2019 this unfortunate person was staring at a backlog of 206 applications – and new ones kept coming every week. The bureaucratic cogs ground to a halt.

Litigious environmentalists did not confine themselves to blocking afforestation applications. They appealed against felling licences and the building of forest roads too. This meant that forest operations almost came to a standstill. Timber was not harvested, so the sawmills and panel board factories had no material to process – they were forced to import logs from Scotland to keep business going. According to Jo O'Hara, a consultant called in by the government to report on the mess, 'during the spring and summer of 2020 the situation developed into a full-blown crisis, which threatened the ongoing operation of the forestry sector as a whole'. Foresters and business leaders were apoplectic, accusing woolly-headed environmentalists of sabotaging a nascent national industry through their 'vexatious' appeals. They formed a new campaigning organisation, took to social media and staged their own protest outside the Dáil, chainsaws and all, to 'prevent the demise of the private forest sector'.

This bureaucratic car crash will probably get worked out, but it points to bigger issues that challenge the Irish forest sector. The rapid afforestation of the twentieth century stimulated powerful opposition from an unlikely alliance of environmentalists, farmers and rural traditionalists. Their guerrilla war against foresters and the forest industry has evolved into something like open battle. The Irish government is stuck in the middle and not sure which way to turn. The consensus that mobilised state afforestation 100 years before has unravelled. At the same time, the need to tackle climate change has made forestry more important than ever. James Moran, an ecologist at the Galway-Mayo Institute of Technology, summed up the situation in 2020: 'It's a sorry state of affairs that in Ireland we are now in the situation that trees are seen as bad for the environment. It's an awful legacy that we've inherited from the previous generation and if we pass on that legacy to the next generation, we have done something seriously wrong.' What is the future for Irish forestry? Is there a way out of this logjam?

A SYLVAN FUTURE?

ON 7 JUNE 2000 A group of ten men convened in the dining room of a farmhouse in County Wicklow to chart a different course for Irish forestry. The meeting was held at Cloragh Farm, a sheep farm and woodland estate near Ashford that has a sideline in hosting film shoots. (Recently, two huge Viking ships loomed up among the trees, part of the set for the TV series *Vikings*.) This was the inaugural meeting of Pro Silva Ireland, an organisation that promotes continuous cover or 'close to nature' forestry. The men studied a letter from Pro Silva Europe welcoming the creation of an Irish branch and inviting them to the next European Congress in Slovakia. A committee was formed, a bank account opened and a newsletter started – all the usual steps for a new charity.

The founding chairperson and driving force behind Pro Silva Ireland was Robert Tottenham. One of the attendees at that first meeting recalled 'we were there for Robert really'. A true pioneer, he was at the forefront of developments in Irish forestry for forty years. Robert was born in 1925 at Mount Callan in County Clare, where his family had farmed since the 1830s. He joined the British army in the Second World War and fought in Europe, India and Burma, before returning home in 1949. Mount Callan is a tabletop mountain only a few kilometres from the Atlantic Ocean, cloaked in heavy, poor soils and battered by wind and rain. For more than a decade Robert used the most modern methods in an attempt to develop his property, but he discovered that farming was not viable on such marginal land. In the late 1960s he met Tom Clear, a forestry professor at University College Dublin, who suggested he plant trees instead. Robert threw himself into this task with gusto, planting 400 hectares over the next twenty years, mostly with Sitka spruce. He developed one of the largest privately owned woods in the country, symbolising the shift from state to private forestry around this time.

Robert was usually seen with an open shirt, a dog whistle around his neck and a springer spaniel at his side. He was remembered by a colleague for 'his mistrust of authority, his love of people, his cheeky smile, his inquisitive eye'. He was always a restless forest owner, never fully satisfied with the advice he received from Irish foresters or the 'tried and trusted' methods. He saw that, as an agricultural country, people in Ireland knew little about true silviculture, so he often looked abroad for inspiration. Influenced by what he read about forestry in South Africa and New Zealand, he started thinning his trees early to increase their resistance to wind. This was at a time when a 'no thin' policy was the norm in the west of Ireland. His trees grew faster and, to the surprise of many, did not blow over.

During a study tour to Germany in the late 1990s, while in his mid-seventies, he first came across a European federation of foresters called Pro Silva, which championed an alternative to clear-felling called 'continuous cover forestry'. The man from Mount Callan had found an alternative to the simplistic rotations back home. According to his youngest brother, when Robert encountered Pro Silva he had what could only be described as a 'near spiritual conversion'. He quickly made his way to the Pro Silva European Congress in Switzerland and wasted no time on his return: Pro Silva Ireland was launched four months later at Cloragh Farm. Robert began to implement continuous cover forestry on his own property on Mount Callan. He was a generous host to students, farmers and forest owners who came to see his methods and persuaded forestry experts from around Europe to visit Ireland and give advice. He was widely mourned when he died at home on Mount Callan in 2007 at the age of eighty-two.

Around half the attendees at the inaugural meeting of Pro Silva Ireland in 2000 were owners of old estates like Robert

Tottenham, some with substantial woodlands. They were inheritors of an Anglo-Irish tradition of estate forestry that had always blended economic, social, environmental and aesthetic values. Despite more than a century of land reform, sectarian conflict and economic decline, they had survived to inject some fresh ideas into the management of Irish forests.

The other co-founders of Pro Silva Ireland were foresters who were dissatisfied with the prevailing model of Irish forestry, often because they had been exposed to different approaches in other parts of Europe. For example, Morgan Roche, the first secretary of the organisation and its most active technical expert during the first decade, had a German mother and had studied forestry at a German university. He bought a run-down forest property near Kilgarvan in County Kerry and applied his new ideas there. (This explains why the first AGM of Pro Silva Ireland was held in Healy Rae's Bar in Kilgarvan, observed by a few curious locals.) Another Irish forester, Paddy Purser, had been exposed to continental methods while spending a summer working in the forests of Switzerland during his studies at University College Dublin. He attended the inaugural meeting of Pro Silva Ireland, was actively involved in the committee for twenty years and served as chairperson from 2016 to 2019. Paddy got involved in the management of Cloragh Forest in County Wicklow and a nearby woodland at Knockrath, introducing continuous cover forestry. He probably has more practical experience than any other Irish forester in this form of silviculture. Another forester with plenty of experience – and Paddy's predecessor as Pro Silva chairperson – was Padraig O'Tuama, a career Coillte forester who took on the lonely task of promoting continuous cover forestry within the state forestry company.

The Pro Silva founders tried to build a broad church. They did their best to entice a wide range of environmentalists, forestry contractors, timber processors, artists and policymakers into their movement. This is epitomised by the journey of another leading member of Pro Silva Ireland – Jan Alexander.

Born in Australia, Jan first visited Ireland in 1979 after some time living in London. 'I loved Irish music and I got a secretarial job in the Australian embassy, which allowed me to stay,' she later recalled. She had a passion for trees but found there was no culture of forestry in Ireland. In 1984 she collected 10,000 acorns and went into schools to teach children how to grow trees from seed. Two years later she co-founded Crann, an environmental non-profit that encouraged broadleaf forestry in Ireland. After she appeared on Gay Byrne's *The Late Late Show*, the most watched programme on Irish television at the time, a flood of new members signed up. Crann launched projects to plant trees and to train people in woodland management around the country.

In the early 1990s, after falling ill, Jan stepped back from the day-to-day running of Crann. This gave her a chance to pause and reflect on what she wanted to do with the rest of her life. The organisation she had co-founded was also changing. Its members were by then mostly urban, and had different priorities. 'Crann had become a tree-lovers' organisation and I was trying to stuff forestry down their throats,' she later said. It was 'less a campaigning organisation' and 'more about encouraging people to appreciate the beauty of trees and how to plant and take care of trees'. Jan wanted to achieve a fundamental shift in the direction of Irish forestry. She was a maverick, similar in personality to Robert Tottenham, always looking for the next frontier as soon as something became mainstream.

In October 2000 she was invited to the first public event organised by Pro Silva Ireland. 'I found myself sitting on the edge of my seat,' she recalled. She was absorbed by a talk by a German guest professor who 'spoke of looking at forestry as a perpetual resource', something that 'humans can benefit from for their own health' as well as being 'an economic activity which also benefits the forest and the environment'. With her typical energy and commitment, Jan embraced Pro Silva Ireland. She served as its chair for a number of years in the mid-2000s. She was a talented public speaker and used her public profile to good advantage.

THE FACT THAT PRO Silva Ireland was a natural home for practical landowners such as Robert Tottenham and environmentalists such as Jan Alexander, along with a smattering of professional foresters, is telling. Normally, Irish forestry suffers from extreme compartmentalisation. We plant 90 per cent conifers here, and 10 per cent broadleaves there, each within its own box. We clear-fell and replant commercial trees, treating them almost like agricultural crops, while being nervous about harvesting timber from any woodland that might be native or ancient. A forest is either a 'wood factory' designed to deliver maximum financial return from selling timber, or a native woodland that supports biodiversity and looks beautiful but with little economic use. A forest is either productive or environmental, never both.

Continuous cover forestry, or 'close to nature' forestry as embraced by Pro Silva Ireland, is a very different approach. It seeks to maintain permanent forest cover and eschews clear-felling. Instead, trees are felled individually or in small groups throughout the entire woodland area. Poor-quality trees (with crooked stems or excessive branching) are removed and high-quality trees are allowed to grow to maturity. The canopy is opened up to let in light and to encourage natural regeneration of new seedlings, which eventually fill the gaps left by felled trees. Over time, a range of tree species naturally emerges across the full breadth of the forest. Although it may start as a monoculture plantation, the end result, after decades or centuries of transformation, is a more natural forest with a diversity of species and age. The overall objective, aptly summarised by one practitioner, is 'to maximise the commercial benefits from an area of woodland while letting natural processes do most of the work'.

Continuous cover forestry may offer a way out of the war that has paralysed the Irish forest sector in recent years. Its philosophical underpinning is something called multifunctional forestry. This is an ugly term for a beautiful concept. The essence is that a forest should deliver multiple economic, social and environmental functions at the same time. Another way to describe it could be 'cake-ism', a phrase popularised by British prime minister Boris Johnson during the Brexit negotiations in 2019 when trying to extract concessions from the EU without giving up anything in return. But, unlike Brexit, continuous cover forestry may be one of those rare examples when we can have our cake and eat it.

Continuous cover forestry can deliver consistent volumes of timber and is entirely compatible with a commercial focus. Research from the UK and Europe shows that this approach can deliver equally good if not slightly better financial returns compared to clear-fell rotations. It is true that a forest owner does not benefit from the low harvesting costs and the lump-sum revenues that come with a big clear-fell event, but this is more than offset by other gains. Under continuous cover forestry the trees are thinned more heavily in the early years, producing earlier cash flow. Trees are allowed to grow to a larger size, which produces a higher proportion of high-value sawlog. Regular harvesting smooths out the impact of fluctuating timber prices and provides steady income. Crucially, successful natural regeneration avoids the costs of replanting, an operation that can swallow up more than one-fifth of the revenue from clear-felling and is a legal obligation in Ireland. Under continuous cover, the 'capital' of the standing forest is retained, while the 'interest' in the form of timber growth is removed every few years. Who needs an investment bond when a forest behaves like this!

Continuous cover forestry can also strengthen resilience to the biophysical threats that may come our way, especially in an era of changing climate. Mixed-aged stands are more resistant to storms, as dominant trees are sturdier and have deeper roots, while the presence of young trees in the understorey lowers wind speeds. And if trees are blown down, the impacts are likely to be less catastrophic as younger trees will already be established and ready to fill the gaps. In contrast, if an even-aged plantation is blown down everything has to be replanted, which sets the rotation back to year zero.

Diverse, mixed-age stands suffer less and recover more quickly from pests and disease. This is logical, as pests and diseases tend to attack specific trees at certain points in their lifecycle. To give some examples from Ireland, the avoidance of clear-felling and reliance on natural regeneration reduces the impact of the large pine weevil, which causes serious damage to conifer seedlings when planted after clear-felling. The latest advice on combating ash dieback is to use continuous cover forestry to develop mixed-species stands, as they will be less vulnerable to disease – it is our only hope to nurse ash trees through their pandemic. And if the worst calamity strikes, and a new disease emerges to attack Sitka spruce, we will want diverse forests rather than the monocultures that currently sustain our timber industries.

Continuous cover forestry has a number of environmental benefits. Forests with a mixture of conifers and broadleaves have healthier soils and do not acidify waterways to the extent of pure spruce stands. Forests managed in this way avoid the release of nutrients and silt into streams that often accompanies clear-felling, with damaging impacts on water quality and sensitive species such as the freshwater pearl mussel. Diverse, mixed-age stands also harbour much more biodiversity than a typical conifer plantation. There are more open spaces, more light reaches the forest floor, there are more species of trees and more 'veteran' trees packed with life at every level, like a New York skyscraper. There is more deadwood lying on the ground, which is an important home for a range of creatures. And there is more time for nature to do its thing. Forest conditions are preserved in perpetuity, which allows species to gradually colonise and compete and adapt, whereas clear-felling wipes out most forest species, setting the ecological clock back to the start each time.

Continuous cover forestry may also provide some of the answer to a carbon accounting conundrum that will face the Irish state over the next thirty years. New forests remove carbon from the atmosphere, but a lot of this carbon is released when a forest is clear-felled at the end of a rotation. This is not a problem if there is an even distribution of forest age across the country, as the growth of young forests will offset the emissions from harvesting. In the case of Ireland, however, this balanced age profile does not exist. And it is getting worse because of the dramatic shortfall in afforestation over the last few years. The forests planted during the peak afforestation of the 1990s and early 2000s are destined to be cleared over the next two decades, but the paucity of recent planting means there will not be enough young forests to offset the carbon emissions. Government figures show that if the current low rate of afforestation continues, forestry will switch from being a net carbon sink to a net source of carbon by 2035, just when Ireland and the EU are striving for carbon neutrality. The timing could not be worse.

One solution to this carbon removal shortfall is to ramp up the level of afforestation to the same levels as before. But this does not seem very likely. The other is to embrace continuous cover forestry. If, instead of clear-felling, we maintain permanent forest cover and selectively harvest the timber growth every few years, most of the carbon will stay locked in the standing trees. In addition, these forests will not experience the loss of carbon from soils that usually accompanies a clear-fell event. There is an opportunity to divert some of the semi-mature plantations planted twenty or thirty years ago from their current path towards clear-felling and instead transform them into mixed-age, diverse, permanent forests. This would smooth the carbon profile of the Irish forest estate and buy more time to reduce emissions from other parts of the economy.

Continuous cover forestry delivers social functions as well. If given a choice, most people would prefer to take a walk in a diverse, mixed-age woodland instead of a dark, dank, mid-rotation Sitka spruce monoculture. Continuous woodlands are more open, more accessible and let in more light. They are more natural, shimmering with an array of plants occupying different layers and niches. They have a better landscape aesthetic. They don't have the sharp lines of a plantation, the solid rectangular blocks of dark green on a hillside that resemble the pixelated computer world of Minecraft. Instead, they blend into the landscape in a natural, irregular way. And neighbours don't have to witness the warzone of chewed branches, stumps and mud left by giant harvesting machines after a clear-fell, a scar on the landscape that can take years to heal. For all these reasons, diverse, mixed-age forests, managed using continuous cover forestry, are much more popular with local communities.

THIS SHIFT IN EMPHASIS from industrial tree-farming to multi-purpose forestry is new in Ireland, but it has been happening in other parts of Europe for decades. Once again, Germany has been at the forefront, and its lessons are relevant to the Irish experience. Recognising this, the founders of Pro Silva Ireland invited one of the leaders of the German continuous cover forestry movement, Professor Hans-Jürgen Otto, to its first public meeting in October 2000. This was the professor who so impressed Jan Alexander with his vision for sustainable forestry. Hans-Jürgen made a number of visits to Ireland with Pro Silva. He was jovial and generous, he never dismissed anyone's opinion, but he was hard to argue with as he had such a deep level of technical knowledge. He also liked the traditions of German forestry, insisting on having lunch in the woods and carrying around a shooting stick that he would unfold and sit on when explaining some point.

Born in Silesia in 1935, Hans-Jürgen studied forestry in Germany and France before joining the State Forestry Administration of Lower Saxony, a large state in north-west Germany that has Hanover as its capital. Lower Saxony had been a pioneer in establishing large areas of spruce plantations on former heathland in the early 1800s (which was similar to the Irish experience in the 1900s, although the Germans used Norway spruce rather than Sitka spruce as the main crop). These even-aged forests were managed to clear-fell and then replanted. Everything went well initially, but by the time of the second or third rotations the system started to fail. The forests became much more vulnerable to storms and started to blow down. To give one traumatic example, in November 1972 a massive storm named Quimburga felled one-tenth of Lower Saxony's forests, or an area of more than 100,000 hectares. The reason was the interaction between the spruce trees and the soil. The weight of mature spruce trees can exceed 600 tonnes per hectare, which places huge pressure on the soil. Compounding this, spruce trees have shallow root systems that fan out around each trunk. As the trees swing in the wind, the root plates lift and fall, stamping down on the soil. This causes compaction of the soil and loss of aeration, which forces the trees to make even shallower roots, compromising stability further. Spruce plantations also produce an acid humus that inhibits the recycling of nutrients and reduces soil fertility. Over time, a spruce monoculture will become less healthy and less stable and more vulnerable to wind damage, pests and diseases.

Hans-Jürgen and his colleagues in the Lower Saxony forestry service also felt increasing pressure from local communities, city-dwellers and new political groups such as the Green Party, who valued forests for recreation and nature rather than timber production. Rather than rejecting these demands, he sought to integrate them into a more holistic approach to forestry. He later articulated his philosophy when accepting the highest forest award in the German-speaking world, the Wilhelm Leopold Pfeil Prize. He believed in an approach that satisfied 'wood production', 'nature conservation' and 'the psychological needs of an increasingly urban population'. He urged his fellow foresters to listen to environmental arguments and to concede past silvicultural mistakes. 'If we negate these ongoing changes in society,' he warned them, 'we as a profession will not survive the next century.'

In the late 1980s Hans-Jürgen led the development of a comprehensive strategy for the long-term ecological development of the state forests of Lower Saxony. This was adopted by the state government in 1991 as the LÖWE programme, and has remained in place ever since. Its central aim is to shift from even-aged spruce plantations to diverse, irregular forests managed using close to nature silviculture. It is a model of 'integrative multifunctional forest management' that has been copied by other German states. Hans-Jürgen became an advocate for this approach across the Continent, co-founding Pro Silva Europe and acting as its president from 1997 to 2000. He travelled all around Europe with his message, which is how he found himself in Ireland in October 2000 talking about the future of Irish forestry.

Continuous cover forestry has continued to gain momentum in Europe since then. Around one-quarter of European forests are now managed in this way. It is required practice in Slovenia, Switzerland and many parts of Germany – the type of clear-felling practised in Ireland is effectively illegal in those areas. In 2002 the Danish government made continuous cover forestry mandatory for all state forests, responding to the increasing incidence of serious wind damage in plantations, coupled with greater public demand for more natural forests. As the climate changes, and new pests and diseases such as the bark beetle afflict stressed trees in central Europe, there is a renewed effort to transform single-species plantations into diverse, mixed-age forests through continuous cover silviculture.

The irony is that Ireland embraced the 'scientific' plantation-based forestry of nineteenth-century Germany just as German foresters were beginning to have second thoughts. Forest management in Germany and other parts of Europe has evolved. The situation in Ireland today feels like Lower Saxony in the 1980s. There are competing demands on Irish forests from industries, environmentalists and recreational users. Opinions are polarised. Active combatants dig deeper trenches around their positions each year. Perhaps the way out of this logjam is a programme of multifunctional forestry built around continuous cover silviculture, a sort of LÖWE programme for Ireland.

WHAT WOULD THIS LOOK like in practice? The first step is to apply continuous cover forestry to existing plantations where this is feasible. Some foresters have the misconception that this method should only be applied to broadleaves or on sensitive ecological sites. In fact, there is a growing body of research on how to use this approach to transform even-aged commercial Sitka spruce plantations. It is important to start early, ideally around the time of the first or second thinning, as then it is possible to thin more heavily without compromising stability and increasing the risk of wind damage. Sitka spruce is a light-demanding species and will not regenerate under a dense canopy, but a skilful forester can open up the canopy through thinning and allow enough light to enter to promote natural regeneration. Once Sitka spruce reaches seed-bearing age, which can happen after thirty years, there can be prolific regeneration of new seedlings. Over time, other species, including native broadleaves, will appear in gaps as well. Supplementary planting can be used to accelerate this process of diversification.

Over many decades – because nature cannot be hurried – what started as an even-aged spruce monoculture will gradually develop into a more diverse forest with intermixtures of conifers and broadleaves. Both types of trees can perform well in this environment; indeed, researchers find that trees grow better in mixed stands rather than in stands of a single type. This makes sense, as each tree species occupies a different vertical layer, is more or less shade-tolerant and accesses different nutrients via its roots. In forestry, two plus two sometimes equals five. This was recognised long ago by Samuel Hayes, the creator of the Avondale estate, who published the first Irish book on tree-planting in 1794. 'Providence has wisely scattered the food of each plant over the surface of the earth', he wrote, 'so that many trees, of different species, will grow well in an acre of ground, where the same number of one kind would actually starve for want of nourishment.'

This brings us to an uncomfortable topic. The biggest threat to the natural regeneration of woodland in Ireland is grazing by wild deer. Deer are one of those species to benefit from the recent expansion of woodland in Ireland and the reduction in sheep numbers on hillsides. Populations of native red deer, sika deer (first imported by Lord Powerscourt in 1860) and fallow deer (introduced by the Anglo-Normans in the twelfth century) have exploded in many parts of the country. Deer eat saplings, rub bark off trees and cause a lot of damage to woodlands, especially broadleaves and native woodland. In a natural ecosystem, deer numbers would be controlled by predators such as wolves or lynx. It is unlikely these animals will be tolerated in Ireland any time soon, so in the meantime humans must take on this role. Ireland lacks a forestry culture, and this means that the sort of hunting culture that can be found in heavily wooded countries such as Germany and France is also less developed. Ireland needs a national deer management strategy, and more active and better-regulated hunters to keep deer numbers under control. To put it crudely, if we want healthy, diverse woodlands, we need to shoot more deer. Ideally, we should develop a taste for venison as well.

Better control of deer will also help with the establishment of new forests in a way that is compatible with future continuous cover management. Beyond transforming existing forests, we should design new plantations so they can evolve towards structural and species diversity. This means planting mixtures of species on the same area, for example groups of broadleaves within a conifer plantation. At the moment, the afforestation grant scheme of the Irish Forestry Programme does not facilitate this. It assumes that species will be compartmentalised, each to its own hectare, and that conifers will always be managed to clear-fell. A reform to this scheme is needed to encourage landowners to plant in a way that will maximise the future benefits of mixed woodland.

What about the choice of species? Do we need to forsake Sitka spruce and the other fast-growing exotic conifers that are the backbone of the Irish forestry industry? The answer is no.

These trees have many benefits: they grow fast; they quickly colonise bare ground, creating rudimentary forest conditions; they produce the timber that our industries need; and they deliver a better financial return for the landowner. Sitka spruce can act as a good pioneer crop, the first step in a much longer transition to a diverse, permanent forest. It can pay the bills over the first few decades while slower-growing species are maturing. This is how Hans-Jürgen Otto viewed the species in an Irish context. Rather than an either/or, Sitka spruce can be an indirect route to mixed forests with a broad range of species. Because this is *multifunctional* forestry, the economic dimension shouldn't be ignored either: a landowner is more likely to take the decision to plant forestry if commercial conifers can be used to deliver an attractive financial return in the first phase.

It is probably time that we dropped our fetishisation of native species. The definition of 'native' is arbitrary and reflects the situation at a single point of time. There is nothing special about the combination of species that managed to recolonise Ireland soon after the last glaciation. As we have seen, there was a much wider range of tree species in Ireland before the last glaciation (including spruces, firs and pines). Since the glaciers disappeared, it is unclear whether certain species died out (such as Scots pine) and whether other species long thought of as alien (such as beech) arrived recently or were here for many thousands of years. And we can expect new species to arrive in the future: the 11,000 years since the last glaciation is a blip in geological time. If plants are well adapted and find a niche, they will establish themselves and become part of the ecosystem.

The environment is constantly changing. Humans are just one more seed-carrying vector in this process, along with the birds or the ocean currents. To see humans as outside of this process, to try to imagine a world without human impact, is another form of anthropocentrism. It is more helpful to situate humans within ecology, shaped by it and capable of shaping it.

The question we should ask, then, is not whether a species is native or exotic. The goal is not to fix an assemblage of species at a certain point in time, or to imagine what the country would look like without humans. Instead, we should ask whether a species, within the right forest management system, can deliver the economic, environmental and social functions that we want from our forests. By this measure, Sitka spruce and other fast-growing conifers can continue to play a role.

Irish people could even learn to love Sitka spruce if it is managed in a new way. In their native range along the Pacific coast of North America, old-growth Sitka spruce forests are wonderful places. In these coastal rainforests, giant trees rise up like mythical giants more than 100 metres high, and there is abundant and diverse understorey vegetation below. These relatively open forests support a great amount of biodiversity: around 350 species of birds and mammals, fifty species of amphibians and reptiles, hundreds of species of fungi and lichens and thousands of species of insects, mites, spiders and other soil organisms. Research from Britain confirms that old-growth Sitka spruce shelters plenty of creatures there too: for example, they are a refuge for the native red squirrel, which has been pushed out of British native broadleaf woodland by another American import, the grey squirrel. A Sitka spruce forest does not have to be a 'green desert'.

The problem is that most Sitka spruce plantations in Ireland are not allowed to grow beyond young adulthood. This is when they are most dreary, and support the least biodiversity. Under continuous cover forestry, the best trees will be allowed to grow longer, into their golden years. We will get more open habitats, more structural complexity and lots of deadwood. In an echo of the great redwood or sequoia forests of North America, huge trees could soar into the sky like the fluted columns of a Gothic cathedral. Given sufficient time, a 'temperate rainforest' could develop in wetter and more sheltered locations. If this happened, Sitka spruce may no longer be reviled as an alien invader, but tolerated and perhaps even welcomed as a strapping immigrant that brings something grand to the Irish landscape.

Continuous cover forestry can't be applied in every situation, however. It is important to recognise its limitations. It is not suitable for older plantations, especially fast-growing ones, that have grown too tall. Introducing heavy thinning at this stage would open the forest up to wind and increase the risk of catastrophic storm damage. In these situations it is better to manage the plantation to clear-fell, using good silvicultural practices to minimise any negative impacts on soils or water. On the next rotation, the land can be planted in a more diverse way with the goal of applying continuous cover forestry from then on.

Continuous cover forestry is also not suitable for wind-prone stands on deep peat soils, no matter what the age. These plantations are too unstable to go through the transformation process. But these are the forests that probably should never have been planted in the first place. They would not receive approval for establishment now. This brings us back to the dirty secret of Irish forestry, that up to one-third of current forests are no good. Continuous cover forestry is not a panacea for this problem. The failed forests of Ireland require their own strategy. The first step is recognition. The Forest Service can't bury its head in the sand and hope a solution emerges: it will require new policies and changes to regulations. Each site will then need its own restoration plan.

There are bog woodland sites with better soils and access that can be profitably felled and then reverted to less demanding native woodland, using species such as birch, alder and Scots pine. Other sites can be clear-felled, but it would be a mistake to attempt to re-establish woodland. Instead, the best option is to block up the drains and let the bogs recover. Coillte have already carried out EU-funded peatland restoration projects like this on small areas, and there are older examples from Britain and other parts of Europe. Peatland restoration can lead to a gradual increase in carbon sequestration as the peat is rewetted, specialist plants grow back and organic carbon is preserved in anaerobic conditions. To make this happen on a larger scale, the Irish Forest Service will need to waive the current obligation to replant forestry after felling. It does not make sense if forests have been established in the wrong place.

There is a final category of peatland forestry that may never be felled, either because it is too inaccessible or the timber is so poor that it won't cover the costs of harvesting. The best option for these sites is to block up the drains and then do nothing, letting nature take its course. The trees may blow over, hardier specimens may survive in stunted form, there may be regeneration, or the bog may slowly swallow everything up – no one really knows. It is rewilding by neglect. It would stand as a reminder to future foresters of the folly of attempting to tame the 'wastes' of Ireland. And in thousands of years a new civilisation might be puzzled to find the fossilised trunks of American lodgepole pine or Sitka spruce lying in the bog, a few layers above much older examples of fossilised Scots pine or oak.

THERE ARE SIGNS THAT policymakers realise that a new approach is needed. In October 2019 legislators passed a motion in Dáil Éireann supporting continuous cover forestry and calling for a move away from clear-felling. In the same year, the Irish Forest Service introduced a new grant scheme that makes extra payments to landowners for implementing continuous cover forestry. Thirty projects, of up to 10 hectares each, were selected. Yet the pace of change is slow. What is needed is a new national strategy that puts multifunctional forestry as its central goal and organises policies and funding around it. Scotland could act as an example. It launched a revised forestry strategy in 2006 around the principle of 'forestry for and with people', embracing the social, economic and environmental functions of forestry, and has seen a revival in afforestation since.

Coillte, the semi-state body that controls half the forest estate in southern Ireland, must be involved. In other countries, such as Germany or Denmark, the state forest sector was the first to embrace ecological forestry because it felt the pressure of public opinion most keenly. In Ireland, the opposite has happened. Pro Silva Ireland was started by private individuals and most of the innovation in continuous cover forestry has been pushed by private foresters. Although individual foresters in Coillte were sympathetic to new ideas, for a long time the organisation continued along the same path, focusing on timber production.

Lately, Coillte has started to change its approach. When the company attempted to clear-fell visible and much-loved forests in places such as the Dublin Mountains there was a public backlash and the company was forced to reconsider. In 2019 Coillte announced a reorganisation. It created a new non-profit entity, Coillte Nature, to focus on 'the restoration, regeneration and rehabilitation of nature across Ireland'. Coillte Nature took over responsibility for the Dublin Mountains estate and developed a new management plan based on continuous cover forestry and planting of native woodland. Coillte has now embraced the use of continuous cover forestry on most of its broadleaf woodlands, ancient woodlands, environmentally sensitive sites and public recreation forests. It aims to increase the area of its forest estate that will be managed primarily for nature to 30 per cent by 2025.

While this is welcome, it could also be another example of the compartmentalisation that afflicts Irish forestry. The intent, it seems, is that Coillte will apply alternative silvicultural methods on a minority of 'sensitive' sites in order to deliver environmental and social functions, while managing its productive conifer forests using conventional methods in order to maximise economic return. If you happen to live beside one of the spruce 'wood factories', or expect a wider range of ecosystem services from them, hard luck. We can only hope that

Coillte will develop expertise in continuous cover forestry that can then be transferred to its full commercial estate.

At the opposite end of the spectrum, the other constituency that will need to be brought along are the more than 20,000 private forest owners, each with an average holding of 8 hectares. Many of these owners lack forestry knowledge. It is expensive to harvest and sell timber from small forests, especially with all the new paperwork, which is why so many are under-managed. Private forests and private forest owners must be brought together to achieve economies of scale and to mobilise the available timber. Those owners who wish to sell will find a ready market created by the new investment funds that have emerged over the last decade to aggregate portfolios. These funds can bring professional management and a long-term perspective. For those who wish to stay in forestry, co-operatives can play an important role. Co-operatives can arrange for joint harvesting and other operations and achieve higher timber prices by aggregating volume. Forestry co-operatives and forest owners' associations play an important part in the vibrant sectors in Denmark and Sweden, where around half the forestry is owned by private individuals, much like in Ireland. There are already active forestry co-operatives in Ireland that can be scaled up, building on the long tradition of successful co-operatives in Irish agriculture.

Perhaps the most important constituency in the future of Irish forestry are all those landowners who have *not* planted trees – and the rural communities around them. Ireland needs more trees, so long as they are the right trees managed in the right way to deliver social, environmental and economic benefits. The world is shifting to a low-carbon bioeconomy that is powered by the energy from today's sun rather than the fossilised solar energy contained in coal, oil and gas. This is the only way we will stave off a climate calamity and bequeath a functioning environment to future generations. Forestry will play a crucial role in this new bioeconomy as a source of materials, a store of carbon, a refuge for biodiversity and an arena for human well-being. Ireland must play its part, growing its own timber for building, manufacturing and energy, rather than living off resources from elsewhere. And it needs to find a way to balance its greenhouse gas emissions from agriculture and other polluting sectors. Afforestation has to become popular again.

Embracing continuous cover forestry may be one solution. A more diverse forest designed to deliver multiple benefits to the landowner and local community is likely to gain more support and to avoid serial objections. It is also more likely to meet the environmental criteria that trip up afforestation applications in sensitive areas. One problem, though, is guaranteeing that continuous cover forestry will be applied in the future and that no future landowner will reverse course and clear-fell. There will need to be novel ways to deal with this. One option is to create a legal mechanism for placing a permanent 'easement' on the land. This is commonly used in the USA, where a landowner agrees not to use land in a certain way, usually for some environmental benefit. The easement is attached to the land, recorded in the land registry, and survives any property sale so that all future owners are bound in the same way. Such a legal mechanism is well suited to forestry as it operates on such long timescales. An easement tied to continuous cover forestry would ensure that a new plantation was gradually transformed into a more diverse, natural woodland while still delivering an economic return.

But perhaps the most important change needed in Ireland is not political or legal, but cultural. We need to develop, or rediscover, a true forest culture. We have been a pastoral people for so long that we have forgotten how to integrate woodlands into our landscapes and society. The revival of interest in fairy trees and sacred trees shows that we retain some links to a past when most of the country was covered in woodland. Yet a true forestry culture will go beyond faux Celtic mysticism. It will appreciate the magic of trees and the spiritual quality of being in woods, but it will also value the usefulness of trees and feel no shame in harvesting a fair share of nature's bounty. We still need wood in much the same way as a thousand years ago: to build our houses, to keep us warm, to make our implements, and to transport our goods. Forests can be beautiful and productive at the same time. If we embrace this culture, we will accept the changes to the Irish landscape as the pendulum swings back from the pastoral to the sylvan. Slowly, Ireland can become an island of woods again.

ACKNOWLEDGEMENTS

This book is the culmination of a long period of research and discussion on the past, present and future of Irish forestry. I owe immense gratitude to Paddy Purser, a pioneering Irish forester who introduced me to the practice of continuous cover forestry and shared innumerable insights in his calm, understated way since we first met in a field in Wicklow in 2014. His colleagues Mark Tarleton and Fionán Russell share a similar ability to combine practical knowledge with an eye for the bigger picture.

I have learnt much from my colleague Darius Sarshar, a forestry professional who has been there and done it across many parts of the world. The rest of my colleagues provide the sort of intellectual sparring partners that keep a mind from going stale. A theme that runs through this book is the importance of transitioning to more ecological forms of land management that address the environmental and social challenges of our era. In our work at SLM Partners, we are lucky enough to be able to turn some of these ideas into action.

Professor Fraser Mitchell of Trinity College Dublin was kind enough to provide comments on the latest thinking on early tree recolonisation and native species in Ireland. I thank all those academics and part-time scholars who have done the hard work of conducting primary research, collecting data and producing the swelling body of research on Ireland's environmental history that this book draws on. The following chapter on sources provides detailed notes on all the publications I used.

I am grateful to my brother Alan for reading through the final manuscript and providing useful comments. My agent Maggie Hanbury was indefatigable in tracking down a publisher that could do this book justice. We found one in New Island:

I would like to thank Aoife K. Walsh, Neil Burkey, Djinn von Noorden and the rest of the publishing team for transforming the manuscript into a finished product.

I thank my family for showing patience and allowing me to escape into the cocoon of writing while in New York during the Covid-19 pandemic. By transporting me back to my homeland, while the sirens rang outside, the experience reminded me that books are the greatest space-shrinking, time-travelling machines ever invented.

SOURCES

This section provides sources for specific information and arguments presented in the text. It follows the order in which material is presented in each chapter.

Introduction

The furore over the Dublin Mountains is in *The Irish Times*, 25 February 2017. The Save Leitrim protest was covered by Agriland on 30 January 2019 (www.agriland.ie/farming-news/trees-are-replacing-people-save-leitrim-protests-outside-dail/) and the forester protest was covered by Agriland on 28 October 2021 (www.agriland.ie/farming-news/failed-and-empty-promises-forestry-group-to-stage-dail-protest).

1. Going Native

Tree colonisation of Ireland is covered in Fraser J. G. Mitchell, 'Where did Ireland's trees come from?', *Biology and Environment: Proceedings of the Royal Irish Academy*, 106B, 3 (2006); Fraser J. G. Mitchell, 'The development of Ireland's tree cover over the millennia', *Irish Forestry*, 58 (2000); and Valerie Hall, 'The history of Irish forests since the Ice Age', *Irish Forestry*, 54 (1997). Michael Viney, *Ireland: A Smithsonian Natural History*, Smithsonian Books (2010) provides a good overview of landscape change – the figure on hazel pollen abundance is from this book. Descriptions of tree species are provided by UK Woodland Trust (www.woodlandtrust.org.uk/) and the Tree Council of Ireland (www.treecouncil.ie). Figures on native plants are from John Parnell and Tom Curtis, *Webb's An Irish Flora*, Cork University Press (2012). Figures on tree genera are from Mitchell, 'The development of Ireland's tree cover

over the millennia'. European vs North American biodiversity is explained in Peter Wohlleben, *The Hidden Life of Trees: What they feel, how they communicate*, William Collins (2017). Pollen analysis is explained in Johnathan Pilcher and Valerie Hall, *Flora Hibernica: The wild flowers, plants and trees of Ireland*, Collins Press (2004) and Peter Woodman, *Ireland's First Settlers: Time and the Mesolithic*, Oxbow Books (2015). Oak genetics are analysed in Colin T. Kelleher et al., 'Irish oak – genetic diversity and the Iberian connection', COFORD Connects (2010). The Iberian origin of species is in Pilcher and Hall (op. cit.). The land-bridge theory is espoused in Mitchell, 'The development of Ireland's tree cover over the millennia' and then debunked in Robin Edwards and Anthony Brooks, 'The island of Ireland: drowning the myth of an Irish land-bridge?', in 'Mind the Gap: Postglacial colonization of Ireland', special supplement to *The Irish Naturalists' Journal* (2008).

The pre-Holocene history of Irish tree species is from Viney (op. cit.). All information on Scots pine is from Jenni Roche et al., 'Plant community ecology of Pinus sylvestris, an extirpated species reintroduced to Ireland', *Biodiversity and Conservation*, 18, 8 (2009) and Jenni Roche, 'Recent findings on the native status and vegetation ecology of Scots pine in Ireland and their implications for forestry policy and management', *Irish Forestry*, 76 (2019). Her findings were confirmed by Alwynne McGeever and Fraser J.G. Mitchell, 'Re-defining the natural range of Scots Pine (*Pinus sylvestris L.*): a newly discovered microrefugium in western Ireland', *Journal of Biogeography*, 43, 11 (2016). The strawberry tree is in Micheline Sheehy Skeffington and Nick Scott, 'Is the Strawberry Tree, Arbutus unedo (*Ericaceae*), native to Ireland, or was it brought by the first copper miners?', *British & Irish Botany*, 3, 4 (2021). Other 'non-native' trees are analysed in Susann Stolze and Thomas Monecke, 'Holocene history of "non-native" trees in Ireland', *Review of Palaeobotany and Palynology*, 244 (2017). Tree adaptation to climate change is described in Wohlleben (op. cit.).

2. Early Humans

The story of the bone discovery is in Marion Dowd, 'A remarkable cave discovery: first evidence for a late Upper Palaeolithic human presence in Ireland', *Archaeology Ireland*, 30, 2 (2016). The most comprehensive treatment of Mesolithic people is found in Peter Woodman, *Ireland's First Settlers: Time and the Mesolithic*, Oxbow Books (2015). Mount Sandel is described in www.irisharchaeology.ie/work/. The story of alder is described in Fraser J. G. Mitchell, 'Where did Ireland's trees come from', *Biology and Environment: Proceedings of the Royal Irish Academy*, 106B, 3 (2006). 'Peak forest' conditions are in articles by Mitchell (op. cit.) and Jonathan Pilcher and Valerie Hall, *Flora Hibernica: The wild flowers, plants and trees of Ireland*, Collins Press (2004). The figure of 85 per cent forest cover is derived from John R. Cross, 'The potential natural vegetation of Ireland', *Biology and Environment: Proceedings of the Royal Irish Academy*, 106B, 2 (2006), who calculates vegetation types based on climate and soil conditions in the past and now. European data is in N. Roberts et al., 'Europe's lost forests: a pollen-based synthesis for the last 11,000 Years', *Nature: Scientific Reports*, 8, 716 (2018).

Neolithic migration is covered in Lara Cassidy et al., 'Neolithic and Bronze Age migration to Ireland and establishment of the insular Atlantic genome', *PNAS*, 113, 2 (2015). Data on archaeological digs is in Woodman (op. cit.) and the M6 digs are analysed in Ellen O'Carroll and Fraser J.G. Mitchell, 'Seeing the woods for the trees: the history of woodlands and wood use revealed from archaeological excavations in the Irish Midlands', *Irish Forestry*, 72 (2015). The story of Irish dendrochronology is in Mike Baillie, 'The radiocarbon calibration from an Irish oak perspective', *Radiocarbon*, 51, 1 (2009). Danish tree-felling is in Eoin Neeson, *A History of Irish Forestry*, The Lilliput Press (1994). Elm decline is covered in Beatrice Ghilardi and Michael O'Connell, 'Fine-resolution pollen-analytical study of Holocene woodland dynamics and land use in north Sligo, Ireland', *Boreas*, 42, 3 (2013) and Valerie Hall, 'The history of Irish forests since the Ice Age', *Irish Forestry*, 54, 1 (1997). Bog formation is from Woodman (op. cit.); Jenni Roche et al., 'Palaeoecological evidence for survival of Scots pine through the Late Holocene in Western Ireland: implications for ecological management', *Forests*, 9, 6 (2018); Michael O'Connell and Karen Molloy, 'Farming and woodland dynamics in Ireland during the Neolithic', *Biology & Environment: Proceedings of the Royal Irish Academy*, 101, 1 (2001); and Beatrice Ghilardi and Michael O'Connell (op. cit.). The demise of Scots pine is explained in Jenni Roche, 'Recent findings on the native status and vegetation ecology of Scots pine in Ireland and their implications for forestry policy and management', *Irish Forestry*, 76 (2019) and Jonathan Pilcher and Valerie Hall, *Flora Hibernica: The wild flowers, plants and trees of Ireland*, Collins Press (2004). For the history of Céide Fields see Michael O'Connell and Karen Molloy (op. cit.) and the most recent research in Michael O'Connell et al., 'Long-term human impact and environmental change in mid-western Ireland, with particular reference to Céide Fields – an overview', *E&G Quaternary Science Journal*, 69, 1 (2020). Wetter conditions are described in Ghilardi and O'Connell (op. cit.). Mitchell quotes are from Fraser J.G. Mitchell, 'The dynamics of Irish postglacial forests', in J.R. Pilcher and Mac an tSaoir (eds), *Wood, Trees and Forests in Ireland*, Royal Irish Academy (1995). 'Boom and bust' is from Nicki J. Whitehouse, 'Neolithic agriculture on the European western frontier: the boom and bust of early farming in Ireland', *Journal of Archaeological Science*, 51 (2014) and British parallels are in O'Connell and Molly (op. cit.). Estimated forest cover is derived from Fraser J.G. Mitchell and Javier Maldonado-Rui, 'Vegetation development in the Glendalough Valley, eastern Ireland over the last 15,000 years', *Biology and Environment: Proceedings of the Royal Irish Academy*, 118B, 2 (2018). Shifts in farming activity are analysed in Nicki Whitehouse et al.,

'Prehistoric land-cover and land-use history in Ireland at 6000 BP', *Past Global Changes Magazine*, 26, 1 (2018).

Bronze Age migration is covered in Lara Cassidy et al., 'Neolithic and Bronze Age migration to Ireland and establishment of the insular Atlantic genome', *PNAS*, 113, 2 (2015). *Fulachtaí fiadha* information is from www.oldeuropeanculture.blogspot.com/2016/07/fulacht-fiadh-cooking-pit.html. Wood use is from Gill Plunkett, 'Land-use patterns and cultural change in the Middle to Late Bronze Age in Ireland: inferences from pollen records', *Irish Quaternary Research*, 18, 4 (2008). The Addergoole canoe is described in www.irisharchaeology.ie/2014/10/the-lurgan-canoe-an-early-bronze-age-boat-from-galway/ and Kelly is quoted in *The Irish Times*, 7 April 2011. Estimating past forest cover is difficult but the 50 per cent estimate is derived from research by Marco Zanon et al., 'European forest cover during the past 12,000 years: a palynological reconstruction based on modern analogs and remote sensing', *Frontiers in Plant Science*, 9 (2018). Associations of yew with Brian Boru and Christianity is from Christine Zucchelli, *Sacred Trees of Ireland*, Collins Press (2016). Yew usage and decline is described in Plunkett (op. cit.) and Pilcher and Hall (op. cit.). Ash development is in O'Carroll and Mitchell (op. cit.) and ash prevalence in hedgerows from in Pilcher and Hall (op. cit.). The appearance of beech, walnut and sycamore is described in Susann Stolze and Thomas Monecke, 'Holocene history of "non-native" trees in Ireland', *Review of Palaeobotany and Palynology*, 244 (2017). The rise of ribwort is referenced in Ellen O'Carroll, 'Ancient woodland use in the Midlands: understanding environmental and landscape change through archaeological and palaeoecological techniques', *Creative minds, Production, Manufacture and Invention in Ancient Ireland*, NRA Monograph Series, 3 (2010) and European developments are presented in N. Roberts et al. (op. cit.). The Bronze Age lull is described by the National Museum of Ireland (www.microsites.museum.

ie/BronzeAgeHandlingBox/bronze-age.html). Climate cooling is from Andrew Bevan et al., 'Holocene fluctuations in human population demonstrate repeated links to food production and climate', *PNAS*, 114, 49 (2017).

3. Celtic Ireland

The Tacitus quote is from Christine Zucchelli, *Sacred Trees of Ireland*, Collins Press (2016). For an overview of Celtic Europe see Peter Bogucki and Pam J. Crabtree, *Ancient Europe 8000 B.C. to A.D. 1000: Encyclopedia of the barbarian world: Vol. 2*, Thomson Gale (2004). Iron Age culture and migrations in Ireland are discussed in J.P. Mallory, *The Origins of the Irish*, Thames and Hudson (2017) and Lara Cassidy et al., 'Neolithic and Bronze Age migration to Ireland and establishment of the insular Atlantic genome', *PNAS*, 113, 2 (2015). *Emain Macha* and Gaelic literature are discussed by Bernard Wailes and Terry Barry in Bogucki and Crabtree (op. cit.).

Early legends and *Craeb Uisnig* are profiled in Alden Watson, 'The king, the poet and the sacred tree', *Études Celtiques*, 18 (1981). The sacred role of trees in Ireland, from pre-Christian times to the present day, is explored in detail in Zucchelli (op. cit.). Much of this section draws on the Zucchelli book.

The disappearance of ceramics is remarked on in Mallory (op. cit.). A good source for trees in early Irish legal texts and literature is Fergus Kelly, 'Trees in early Ireland', *Irish Forestry*, 56 (1999). This is also covered in Eoin Neeson, *A History of Irish Forestry*, The Lilliput Press (1994). The burning of oak is explored in Lorna O'Donnell, 'Into the woods: revealing Ireland's Iron Age woodlands through archaeological charcoal analysis', *Environmental Archaeology*, 23 (2018). The Annals of the Four Masters are described in Zucchelli (op. cit.). Details on staves can be found in Martin G. Comey, 'Stave-built wooden vessels from Medieval Ireland', *The Journal of*

Irish Archaeology, 12/13 (2003–4). Chariots are discussed in Raimund Karl, 'Iron Age chariots and medieval texts: a step too far in "breaking down boundaries"?', *e-Keltoi: Journal of Interdisciplinary Celtic Studies*, 5 (2003), while Valerie Hall, *The Making of Ireland's Landscape Since the Ice Age*, Collins Press (2011) describes the Corlea Bog trackway. The section on the ownership and management of woods draws from Kelly (op. cit.) and Aidan O'Sullivan, 'The use of trees and woodland in early medieval Ireland', *Irish Forestry*, 51 (1994), and to a lesser extent Nigel Everett, *The Woods of Ireland: A history, 700–1800*, Four Courts Press (2014).

Ferriter's Cove and early cattle farming are covered in Michael O'Connell et al. (eds), *Cattle in Ancient and Modern Ireland: Farming practices, environment and economy*, Cambridge Scholars Publishing (2016) and Hall (op. cit.). Roman perceptions are quoted in Mallory (op. cit.). Woodland decline is tracked in Beatrice Ghilardi and Michael O'Connell, 'Fine-resolution pollen-analytical study of Holocene woodland dynamics and land use in north Sligo, Ireland', *Boreas*, 42, 3 (2013) and the more scrub-like environment in Ellen O'Carroll and Fraser J.G. Mitchell, 'Seeing the woods for the trees: the history of woodlands and wood use revealed from archaeological excavations in the Irish Midlands', *Irish Forestry*, 72 (2015). Wetter conditions are noted by O'Donnell (op. cit.). Clonycavan Man and trackways are described in Hall (op. cit.) and O'Carroll and Mitchell (op. cit.). The revival of woodland during the Late Iron Age lull is explored in Hall (op. cit.) and O'Donnell (op. cit.). Roman migrants are noted by Mallory (op. cit.). The situation in Europe is explored in N. Roberts et al., 'Europe's lost forests: a pollen-based synthesis for the last 11,000 Years', *Nature: Scientific Reports*, 8, 716 (2018). The increase in agriculture is detected by Anette Overland and Michael O'Connell, 'New insights into late Holocene farming and woodland dynamics in western Ireland with particular reference to the early medieval horizontal watermill at Kilbegly, Co. Roscommon', *Review of*

Palaeobotany and Palynology, 163 (2011). The ninth-century list is from Kelly (op. cit.). Iria the Prophet is mentioned in Neeson (op. cit.). Forest cover estimates are based on data in Ralph Fyfe et al., 'The Holocene vegetation cover of Britain and Ireland: overcoming problems of scale and discerning patterns of openness', *Quaternary Science Reviews*, 73 (2013) and Ralph Fyfe et al., 'Deforesting Europe: pollen-inferred Holocene land cover change using the "pseudobiomisation" approach', *PANGAEA* (2015).

4. Medieval Invaders

Information on *Skuldelev 2* is from the Viking Ship Museum (www.vikingeskibsmuseet.dk/en/visit-the-museum/exhibitions/the-five-viking-ships). The age of oak used is from Eileen Reilly et al., 'Building the towns: the interrelationship between woodland history and urban life in Viking Age Ireland', in Ben Jervis et al. (eds), *Objects, Environment, and Everyday Life in Medieval Europe*, Turnhout (2016).

The Viking period is summarised in Alex Woolf, 'The Scandinavian intervention', in Brendan Smith (ed.), *The Cambridge History of Ireland: Volume I*, Cambridge University Press (2018). First coins are referenced in Margaret Murphy, 'The economy', in Smith (ed.) (op. cit.), as are Viking trade and the commercial role of Dublin. Timber exports to Iceland/Faroes are described in Eoin Neeson, *A History of Irish Forestry*, The Lilliput Press (1994). The details on wood use in Viking Dublin (and Waterford) are found in Reilly et al. (op. cit.) and Patrick Wallace, 'Viking Dublin', in Peter Bogucki and Pam J. Crabtree, *Ancient Europe 8000 BC–AD 1000: Encyclopedia of the barbarian world: Vol. 2*. The 'Viking lake' is from Wallace (op. cit.). The population figure for Dublin is taken from Murphy (op. cit.). Neeson (op. cit.) refers to the bridge in Dublin. Wallace (op. cit.) describes the maritime power of Dublin. The sons of King Harold II are described by the Viking Ship Museum (www.

vikingeskibsmuseet.dk/en/professions/education/viking-knowl-edge/the-viking-age-geography/the-vikings-in-the-west/england/the-battle-for-england) and Nigel Everett, *The Woods of Ireland: A history, 700–1800*, Four Courts Press (2014).

The Anglo-Norman invasion is described in Colin Veach, 'Conquest and Conquerors', in Smith (ed.) (op. cit.). Quotes are from Giraldus Cambrensis, *The Topography of Ireland*, various editions. Geoffrey of Monmouth is quoted in Nicholas Vincent, 'Angevin Ireland', in Smith (ed.) (op. cit.). English woodland cover is estimated in Oliver Rackham, *The History of the Countryside*, Weidenfeld and Nicolson (1986). The second Gerald of Wales quote is from Giraldus Cambrensis, *The Conquest of Ireland*. See also Neeson (op. cit.). Irish military tactics are analysed in Veach (op. cit.). Julius Caesar is quoted in Nigel Everett (op. cit.), which also describes the 1297 Act and the activities of Art MacMurrough Kavanagh. The manorial system is profiled in Murphy (op. cit.) and parkland in Fiona Beglane, *Anglo-Norman Parks in Medieval Ireland*, Four Courts Press (2015). Woodland management is covered in Peadar Slattery, 'Woodland management, timber and wood production, and trade in Anglo-Norman Ireland, c.1170 to c.1350', *The Journal of the Royal Society of Antiquaries of Ireland*, 139 (2009). The expansion of towns is in Murphy (op. cit.). Henry II's court is described in Vincent (op. cit.). Reilly (op. cit.) outlines timber construction techniques. The use of moss is from Aidan O'Sullivan, 'The use of trees and woodland in early medieval Ireland', *Irish Forestry*, 51 (1994). Drogheda's struggles are in Murphy (op. cit.), St Patrick's Cathedral is in Slattery (op. cit.), and the ship order is described in Everett (op. cit.). Rod-cutting and timber exports are explored in Slattery (op. cit.). Economic and demographic expansion are described in Murphy (op. cit.). The figure of 20 per cent forest cover is estimated by R.E. Glasscock, 'Land and people, c.1300', in Art Cosgrove (ed.), *A New History of Ireland, Volume II: Medieval Ireland 1169–1534*, Oxford University Press (2008).

Famine, plague, climate change and woodland revival are found in Valerie Hall, *The Making of Ireland's Landscape Since the Ice Age*, Collins Press (2011) and Murphy (op. cit.). The expansion of hazel scrub is recorded in Ellen O'Carroll and Fraser J.G. Mitchell, 'Seeing the woods for the trees: the history of woodlands and wood use revealed from archaeological excavations in the Irish Midlands', *Irish Forestry*, 72 (2015). The story of oak tree rings is in Hall (op. cit.). The decline of English authority is in Christopher Maginn, 'Continuity and change 1470–1550', in Smith (ed.) (op. cit.); the Gaelic cultural revival appears in Brendan Smith, 'Disaster and Opportunity, 1320–1450', in Smith (ed.) (op. cit.); and the emergence of *creaght* and tower houses is in Murphy (op. cit.). The population figure is from S.J. Connolly (ed.), *The Oxford Companion to Irish History*, Oxford University Press (1999). Gaelic trade is described in Maginn (op. cit.) and Everett (op. cit.). The quote 'air of fraud' is from Everett (op. cit.). Links with Spain are described in Raymond Gillespie, 'Economic life, 1550–1730', in Jane Ohlmeyer (ed.), *The Cambridge History of Ireland: Volume II*, Cambridge University Press (2017). The quote '*Teóra h-uaire*' is from Kenneth Nicholls, 'Woodland cover in pre-modern Ireland', in P.J. Duffy et al. (eds), *Gaelic Ireland c.1250–c.1650: Land, lordship and settlement*, Four Courts Press (2001).

5. Conquest and Commerce

Tomm Moore and Ross Stewart (dir.), *Wolkwalkers*, Cartoon Saloon (2020). The film review is in *The Guardian*, 30 October 2020. The quote 'most rapid transformation' is from Jane Ohlmeyer, 'Introduction', in Jane Ohlmeyer (ed.), *The Cambridge History of Ireland: Volume II*, Cambridge University Press (2017).

The myth of dense Irish forest cover is still being peddled today. See J. Ruby Harris-Gavin, 'Ireland's forest fallacy', *Éire-Ireland*, 55, 3&4 (2020) which states that 'Ireland remained largely forested without major human intervention until 1600'. No one who has seriously studied the subject believes this anymore. But even among serious scholars there are differences of opinion on the extent of forest cover at this time. Eileen McCracken in her book *The Irish Woods Since Tudor Times: Their distribution and exploitation*, David and Charles (1971) was brave enough to estimate total forest cover in 1600 at 12.5 per cent based on her study of contemporary maps and texts. This was broadly accepted until the English ecologist Oliver Rackham made his own study of the Civil Survey maps of 1654–6 and concluded that less than 3 per cent of the island was wooded at that time – see Oliver Rackham, *The History of the Countryside*, Weidenfeld and Nicolson (1986). Both the 12.5 per cent and 3 per cent figures have appeared in studies since then, including in publications by Irish governmental departments. Eoin Neeson in his *A History of Irish Forestry*, The Lilliput Press (1994) sides with McCracken, emphasising the amount of woodland destruction that accompanied the English conquest. Valerie Hall in her *The Making of Ireland's Landscape Since the Ice Age*, Collins Press (2011) sits on the fence, presenting evidence to support both views – 'the fate of the Irish woodlands' quote is from this book. Nigel Everett in his *The Woods of Ireland: A history, 700–1800*, Four Courts Press (2014) does not venture a quantitative estimate but appears sympathetic to Rackham and downplays the size of Irish woodlands at the time of the English conquest. However, the most comprehensive study of sixteenth- and seventeenth-century Irish maps can be found in William J. Smyth, *Map-making, Landscapes and Memory: A geography of colonial and early modern Ireland* c.*1530–1750*, University of Notre Dame Press (2006). Smyth comes down firmly on the side of McCracken. Rather than relying only on the Civil Survey, he analyses earlier, local maps (for example, for Carlow and Ulster) that show high levels of woodland cover. It is puzzling how Rackham, an undoubted master at reading the history of the British countryside, could come to such a different conclusion to Irish experts. It could be that he only dabbled in Ireland and did not interpret the sources correctly. This author is of the opinion that the balance of evidence points to a forest cover in the late sixteenth century of around 10–15 per cent, which was McCracken's original assessment.

The description of major Irish woodlands is from McCracken (op. cit.) and Kenneth Nicholls, 'Woodland cover in pre-modern Ireland', in P.J. Duffy et al. (eds), *Gaelic Ireland c.1250–c.1650: land, lordship and settlement,* Four Courts Press (2001).

For O'Neill see Roy Foster, *Modern Ireland: 1600–1972*, Penguin Books (1990) and *The Oxford Companion to Irish History*. Military tactics and observations of Glenkonkyne wood are from Everett (op. cit.), although the quote 'the woods and bogs' is from Hall (op. cit.). Shakespeare references are from Willy Maley and Rory Loughnane (eds), *Celtic Shakespeare: the Bard and the Borderers*, Ashgate Publishing (2013). Figures on the exploitation of Glenkonkeyne and Killetra are from Neeson (op. cit.) and Everett (op. cit.). The London government reaction is described in Hall (op. cit.). Pollen analysis from this area is in Valerie Hall, 'Pollen analytical investigations of the Irish landscape AD 200–1650', *Peritia*, 14 (2000). The 'so spoiled' quote is from Everett (op. cit.). The removal of hazel scrub and end of the silvo-pastoral tradition is explored in Hall, *The Making of Ireland's Landscape Since the Ice Age*. The treatment of this type of land in maps is in Nicholls (op. cit.).

The Desmond wars are described in Everett (op. cit.) and *The Oxford Companion to Irish History*. Raleigh is described in Everett (op. cit.). The East India Company details come from Everett (op. cit.) and Joseph Nunan, 'Boyle and the East India Company in Co. Cork: a case study in colonial competition', in David Edwards and Colin Rynne (eds), *The Colonial World of Richard Boyle, First Earl of Cork*, Four Courts Press (2018). The quotes 'model planter', 'pursuer of Irish land' and 'tycoon in a colonial setting' are from David Edwards and Colin Rynne, 'Introduction', in Edwards and Rynne (ed.) (op. cit.). The quote 'poster boy' is from Colin Rynne, 'Colonial entrepreneur and urban developer' in the same book. All other details on Boyle are from articles in this recent book. The number of ironworks in Ireland is in McCracken (op. cit.). The quote 'perhaps the bloodiest' is from Micheál Ó Siochrú and David Brown, 'The Down Survey and Cromwellian Land Settlement' in Ohlmeyer (ed.) (op. cit.).

Crown attempts to regulate timber trade are explored in Everett (op. cit.). The industrial revolution paragraph is derived from McCracken (op. cit.). Changes in land ownership are from Raymond Gillespie, 'Economic life, 1550–1730', in Ohlmeyer (ed.) (op. cit.). Rackham's comments on Irish ironworks are in Oliver Rackham, 'Looking for ancient woodland in Ireland', in J.R. Pilcher and Mac an tSaoir (eds), *Wood, Trees and Forests in Ireland*, Royal Irish Academy (1995) and his quotes on tree clearance are from Oliver Rackham, *The History of the Countryside*, Weidenfeld and Nicolson (1986). The Gillespie quote is from Gillespie, 'Economic life, 1550–1730', in Ohlmeyer (ed.) (op. cit.). Boyle's finances are analysed in Colin Rynne (op. cit.). Population figures and the size of Dublin are from Gillespie (op. cit.). Immigration per cent is from Ohlmeyer (op. cit.). The figures on agricultural development and trade are from Gillespie (op. cit.) and Francis Ludlow and Arlene Crampsie, 'Environmental history of Ireland, 1550–1730', in Ohlmeyer (ed.) (op. cit.). Timber needs and net imports are explored in McCracken (op. cit.). The average age of oak trees is from Ludlow and Crampsie (op. cit.). Boate is quoted in S. Mendyk, 'Gerard Boate and "Irelands Naturall History"', *The Journal of the Royal Society of Antiquaries of Ireland*, 115 (1985) and Everett (op. cit.).

6. The Two Irelands

Biographical details on Burke are from Sean Donlan, 'Edmund Burke: scorned loyalty and rejected allegiance', *History Ireland*,

2, 12 (2004). Burke quotes are in Edmund Buke, 'Letter to Sir Hercules Langrishe' (1792), in Philip Magnus (ed.), *Prose of Edmund Burke*, Falcon Prose Classics (1948). Penal laws are covered in D.W. Hayton, 'The emergence of a Protestant society, 1691–1730', in Jane Ohlmeyer (ed.), *The Cambridge History of Ireland: Volume II*, Cambridge University Press (2017).

For land security and tree-planting see Eileen McCracken, *The Irish Woods Since Tudor Times: Their distribution and exploitation*, David and Charles (1971). Big Houses and demesnes are explored in Terence Reeves-Smyth, 'The natural history of demesnes', in John Wilson Foster and Helena C.G. Chesney (eds), *Nature in Ireland: A scientific and cultural history*, The Lilliput Press (1997) and Michael Viney, *Ireland: A Smithsonian natural history*, Smithsonian Books (2010). Hedgerows are described in detail by Viney (op.cit.). Length of hedgerows is from Teagasc, 7 December 2020 (www.teagasc.ie/news--events/daily/environment/mapping-irish-hedgerows.php). The 4 per cent figure is from Kevin Black et al., *Carbon Sequestration by Hedgerows in the Irish Landscape*, Environmental Protection Agency, CCRP Report No. 32 (2014). Timber use, timber prices, regulations and timber import data are from McCracken (op. cit.). Bog wood is mentioned in Andy Bielenberg, 'The Irish economy, 1815–1880', in James Kelly (ed.), *The Cambridge History of Ireland: Volume III*, Cambridge University Press (2018). Marshall quote and tanning needs are contained in Nigel Everett, *The Woods of Ireland: A history, 700–1800*, Four Courts Press (2014). The Swift quote is from Jonathan Swift, *Drapier's Letters*, Letter VII (1724), Project Gutenberg. The history of legislation is given in Everett (op. cit.) and McCracken (op. cit.). The quote 'the principal agent' is from David Dickson, 'Society and Economy in the Long Eighteenth Century', in Kelly (ed.) (op. cit.). Madden quote and 1783 scheme are from Everett (op. cit.). Premiums and species planted are from McCracken (op. cit.). Sycamore and beech are described in Johnathan Pilcher and Valerie Hall, *Flora Hibernica: The wild flowers, plants and trees of Ireland*, Collins Press (2004), while Peter Wohlleben, *The Hidden Life of Trees*, William Collins (2017) describes European dominance of beech. Information on species is also given by UK Woodland Trust. Ellis and Smith quotes are in Everett (op. cit.). Planting figures and Registrar quotes are from Eoin Neeson, *A History of Irish Forestry*, The Lilliput Press (1994). Sale and clearing of woodland is described in McCracken (op. cit.). Shillelagh woods history is given in Everett (op. cit.) and also Kenneth Nicholls, 'Woodland cover in pre-modern Ireland', in P.J. Duffy et al. (eds), *Gaelic Ireland c.1250–c.1650: Land, lordship and settlement*, Four Courts Press (2001). Modern events are from *The Irish Times*, 28 Apr 2011. Young quote is from McCracken (op. cit.).

Forest areas in this chapter are derived from Irish Census figures from 1791 to 1881, as reproduced in William J. Smyth, 'The greening of Ireland: tenant tree-planting in the eighteenth and nineteenth centuries', *Irish Forestry*, 54, 1 (1997); McCracken (op. cit.); Neeson (op. cit.); Bielenberg (op. cit.); and Niall O'Carroll, *The Forests of Ireland*, Turoe Press (1984). A word of caution. The census figures sometimes refer to forest plantations only, and sometimes to plantations plus orchards and hedgerows. In most cases, the census-takers did not record 'natural woods', i.e. unplanted areas, so the census figures probably underestimate the total area of woodland. Not all authors pay attention to these different categories when quoting the figures, so there can be discrepancies. UK forest cover is from Bielenberg (ed.) and German data is from Baron von Viettinghoff-Riesch, 'Outlines of the history of German forestry', *Irish Forestry*, 15 (1958).

The Lewis quote is from Anna Pilz and Andrew Tierney, 'Trees, Big House culture, and the Irish literary revival', *New Hibernia Review*, 19, 2 (2015) and the second traveller quote is from Philippe Daryl [Paschal Grousset], *Ireland's Disease; Notes and impressions*, George Routledge and Sons (1888). Rural poverty is described in David Dickson, 'Society and economy in the long eighteenth century', in James Kelly (ed.), *The Cambridge History of Ireland: Volume III*, Cambridge University Press (2018). The Young quote is from Everett (op. cit.), as is the following section on agrarian violence and Gaelic hostility to estates. Liberty Tree is referenced in Christine Zucchelli, *Sacred Trees of Ireland*, Collins Press (2016) and ash tree rumours in Everett (op. cit.). The population figure is from Brian Gurrin, 'Population and emigration, 1730–1845', in Kelly (ed.) (op. cit.), and population density from Peter Gray, 'The Great Famine, 1845–1850', in Kelly (ed.) (op. cit.). Dependence on potato and expansion of agriculture is in Bielenberg (op. cit.) and Dickson (op. cit.). Lake island information is from this author's interview with Fraser J.G. Mitchell (2021). Aran Islands are profiled in Karen Molloy and Michael O'Connell, 'Fresh insights into long-term environmental change on the Aran Islands', *Journal of the Galway Archaeological and Historical Society*, 59 (2007). The famine is explored in Gray (op. cit.). Post-famine livestock expansion is from Bielenberg (op. cit.) and Viney (op. cit.).

Land ownership data is from Paul Bew, *Ireland: The politics of enmity 1789–2006*, Oxford University Press (2007). The McCracken quote is from McCracken (op. cit.) and Forbes is quoted in Pilz and Tierney (op. cit.). Clearing of woodlands at this time is also analysed by Niall O'Carroll in *The Forests of Ireland*, Turoe Press (1984) and *Forestry in Ireland – a concise history*, COFORD (2004). 1907 data is from the Departmental Committee 1908b, cited in O'Carroll, *Forestry in Ireland – a concise history*. First World War clearing is described in A.C. Forbes, 'Tree planting in Ireland during four centuries', *Proceedings of the Royal Irish Academy: Archaeology, Culture, History, Literature*, 41 (1932–4). 1922 figures are from Department of Agriculture and Technical Instruction for Ireland (DATI) reports (1923–5) cited in Neeson (op. cit.). Taboos around fairy trees are explored in Zucchelli (op. cit.).

7. Reforesting Ireland

The 'Cradle' quote is from *Avondale House and Forest Park; a guide to the Forest Park*, Coillte (2006). Avondale history is in Niall O'Carroll, *Forestry in Ireland – a concise history*, COFORD (2004). Redevelopment is profiled in the *Irish Independent*, 5 June 2021, and on the German company website (www.eak-ag. de/eakag-en/about-us/who-we-are/).

'As treeless as Portugal' quote is from James Joyce, *Ulysses*, Penguin (2000). Hogan quote is from Dáil Debate, 3 May 1928, cited in Margaret Duff Garvey, 'Replanting Ireland: parliamentary debate and expert literature on Irish state forestry 1922 to 1939', PhD Dissertation, Trinity College Dublin (2020). Howitz is covered in A.C. Forbes, 'Some early economic and other developments in Eire, and their effect on forestry conditions', *Irish Forestry*, 1 (1943) and Eoin Neeson, *A History of Irish Forestry*, The Lilliput Press (1994). The Knockboy fiasco is in Litton C. Falkiner, 'The forestry question considered historically', *Journal of the Statistical and Social Inquiry Society of Ireland*, 11 (1902/1903); H.M. FitzPatrick, *The Forests of Ireland*, Society of Irish Foresters (1966); and Neeson (op. cit.). Forbes description is by Edward Lysaght, quoted in Niall O'Carroll (op. cit.). Quote 'father of Irish forestry' is from Margaret Duff Garvey (op. cit.). The Departmental Committee is analysed by O'Carroll (op. cit.) and Falkiner (op. cit.), while Neeson (op. cit.) makes the comparison with British forestry. Details of Forbes's work, 'the future of woods' quote and 1928 Act are from Neeson (op. cit.). Dillon quote is from O'Carroll (op. cit.). Castle Durrow is from O'Carroll (op. cit.) and www.castledurrow.com/castle-durrow-hotel-history.html. Quote 'poorest off' is from Roy Cameron, *Report on forestry mission to Ireland* (1951), quoted in Neeson (op. cit.). 1907 forestry statistics are from Departmental Committee 1908b, cited in O'Carroll (op. cit.). MacBride is profiled in *The Oxford Companion to Irish History*, Oxford University

Press (1999), his quote 'keenly interested' is from O'Carroll (op. cit.) and his role is also described in Neeson (op. cit.). Whitaker and government policy are explained in Neeson (op. cit.) and O'Carroll (op. cit.). Quote 'gradual lowering' is from H.J. Gray, 'The economics of Irish forestry', *Journal of the Statistical and Social Enquiry Society of Ireland*, 21, 2 (1963). State afforestation in Ireland, north and south, is detailed in O'Carroll (op. cit.).

'Scientific' forestry is analysed in Joachim Radkau, 'Wood and forestry in German history: In quest of an environmental approach', *Environment and History*, Vol. 2, No. 1 (1996) and Peter Savill, 'High forest management and the rise of even-aged stands', in K.J. Kirby et al. (eds), *Europe's Changing Woods and Forests: From wildwood to managed landscapes*, CABI (2015). Reinhard's colourful story is in David O'Donoghue, 'The story of Otto Reinhard: a case study of divided loyalties in peace and war', *Irish Forestry*, 69 (2012). Douglas fir, lodgepole pine and Sitka spruce are described in O'Carroll, *The Forests of Ireland*. Areas for each species are derived from DAFM, *Ireland's national forest inventory 2017*. Sitka afforestation data is from DAFM, *Forest Statistics Ireland 2020*. Numbers of Sitka spruce trees in UK and Ireland is from Ruth Tittensor, *Shades of Green: An environmental and cultural history of Sitka spruce*, Windgather Press (2016).

Curtailment of state forestry is described in Neeson (op. cit.), which also contains the McEvoy quote. Sidelining of Anglo-Irish landlords is described in Garvey (op. cit.). Early grant schemes are explored in O'Carroll, *Forestry in Ireland* and Mary Ryan et al., 'The role of subsidy payments in the uptake of forestry by the typical cattle farmer in Ireland from 1984 to 2012', *Irish Forestry*, 71 (2014). Investor activity is revealed in Tim O'Brien, 'Private forestry in Ireland: Recent achievements and future direction', *Irish Forestry*, 48 (1991). Coillte's history is given in O'Carroll, *Forestry in Ireland* and its loss of grant support in the *Irish Independent*, 26 November

2003. Comparisons of Ireland to European forestry draw on Forest Europe, *State of Europe's Forests 2015*. Information on 1980s forest ownership is from Niall O'Carroll, *The Forests of Ireland*. Current ownership is from DAFM, *Forest Statistics Ireland 2020*, as are figures on net exports. Reappearance of woodpeckers is described in www.irelandswildlife.com/great-spotted-woodpecker-ireland.

All data in this and future chapters on afforestation in the Republic of Ireland is taken from the Irish government's *2021 Afforestation Statistics* (www.gov.ie/en/collection/15b56-forest-statistics-and-mapping/#afforestation-statistics), which has data back to 1921. Data on Northern Ireland is from Forest Research (www.forestresearch.gov.uk/tools-and-resources/statistics/data-downloads/), although this dataset only goes back to 1976.

This author has compiled data from different sources to estimate total forest cover for the island of Ireland at different points over the last 120 years. Inaccurate figures can be found in other sources, as some quote figures for the Republic of Ireland as if they apply to the whole island. Data from the Republic and Northern Ireland must be merged to give an all-island picture. A summary table is given on the next page and these figures are quoted in the main text.

The 1907 data is from Departmental Committee 1908b, cited in O'Carroll, *Forestry in Ireland*. The 1925 data for the Republic is from Department of Agriculture and Technical Instruction for Ireland (DATI) Reports (1923–5) cited in Neeson (op. cit.) and for Northern Ireland is from J.R. Aldhous, 'British forestry: 70 years of achievement', *Forestry: An International Journal of Forest Research*, 70, 4 (1997). 1950 data for the Republic is from Roy Cameron, *Report on forestry mission to Ireland* (1951), quoted in Neeson (op. cit.) and for Northern Ireland is from Forest Research, *Forestry Statistics 2021*. The 1965 data for the Republic is from *Annual Report of the Department of Agriculture 1964/65*, cited in DAFM, *Forestry Statistics Ireland 2021* and for Northern Ireland is from Forest Research, *Forestry Statistics 2021*.

Year	Republic	North	All Ireland	Per cent forest cover
1907	102,627	16,790	119,417	1.4 per cent
1925	100,567	18,000	118,567	1.4 per cent
1950	98,075	23,448	121,523	1.5 per cent
1965	254,350	42,758	297,108	3.6 per cent
1973	323,654	56,667	380,321	4.6 per cent
1980	375,000	66,000	441,000	5.3 per cent
2000	649,918	83,000	732,918	8.8 per cent
2006	697,730	90,000	787,730	9.4 per cent
2012	731,650	110,000	841,650	10.1 per cent
2017	770,020	112,000	882,020	10.6 per cent
2020	780,030	118,000	898,030	10.7 per cent

The 1973 data for Republic of Ireland is from T.J. Purcell, *Inventory of private forests – 1973*, Department of Fisheries and Forestry and from *Annual Report of the Department of Agriculture 1972/73* cited in DAFM, *Forestry Statistics Ireland 2021*. 1973 data for Northern Ireland is an estimate based on data for 1970 cited in House of Lords Debate, 3 February 1988, *Hansard*, vol. 492, cc1157–80 and data for 1979 from Northern Ireland Forest Services, *Progress Report 1973–79*. The 1980 data for Republic of Ireland is based on government estimates in Niall O'Carroll, *The Forests of Ireland*, Turoe Press (1984), which are undated but probably refer to the period around 1980. The 1980 data for Northern Ireland is based on the 1979 estimate in Northern Ireland Forest Services, *Progress Report 1973–79*. 2000 data for the Republic is from COFORD, *Forecast of roundwood production from the forests of Ireland 2001–2015* and for Northern Ireland is from Forest Research, *Forestry Statistics 2021*. Data for the Republic for 2006, 2012 and 2017 is from the *National Forest Inventory 2007, 2013 and 2017*. The 2020 data for the Republic is based on adding the subsequent afforestation area to the 2017 forest area in the *National Forest Inventory 2017*. Northern Ireland data for 2006, 2012, 2017 and 2020 is from Forest Research, *Forestry Statistics 2021*.

8. Grinding to a Halt

Farrelly research is in Niall Farrelly and Gerhardt Gallagher, 'The potential availability of land for afforestation in the Republic of Ireland', *Irish Forestry*, 72 (2015). Sawmill investment is in Cathal Geoghegan et al., 'The Irish forestry sector' and in Cathal O'Donoghue et al. (eds), *Rural Economic Development in Ireland*, Teagasc (2014). Export values are given in DAFM, *Forest Statistics Ireland 2020*. Per capita emissions are presented in Carbon Brief, 18 June 2019 (www.carbonbrief.org/in-depth-qa-why-ireland-is-nowhere-near-meeting-its-climate-change-goals). Pulpwood prices are in ITGA, *Forestry and timber yearbook 2016*. Policy goals are in the Irish government's *Growing for the Future: a strategic plan for the development of the forestry sector in Ireland* (1996). A bullish perspective can be found in Tim O'Brien, 'Private forestry in Ireland: recent developments and future directions', *Irish Forestry*, 48 (1991).

The influence of subsidies is described in Siobhan McCarthy, 'The future of the forestry sector in Ireland', *Irish Banking Review*, Winter (2002). Farmer attitudes and the first two quotes are from Mary Ryan, 'The role of subsidy payments in the uptake of forestry by the typical cattle farmer in Ireland from 1984 to 2012', *Irish Forestry* (2014) and 'Socio-economic drivers of farm afforestation decision-making', *Irish Forestry*, 73 (2016). Idea of loss is from John McDonagh et al., 'New opportunities and cautionary steps? Farmers, forestry and rural development in Ireland', *European Countryside*, 4 (2010). Reluctance to plant is analysed in Peter Howley et al., 'Farm and farmer characteristics affecting the decision to plant forests in Ireland', *Irish Forestry*, 69 (2012). Opposition to forestry is described in Áine Ní Dhubháin, 'The impact of forestry on rural communities', *Irish Forestry*, 52 (1995); Roy Tomlinson and John Fennessy, 'Sustainable forestry in Northern Ireland and the Republic of Ireland', in John McDonagh et al. (eds), *A Living Countryside?*, Routledge (2009); Matthew Carroll et al., 'Afforestation and local residents in County Kerry,

Ireland', *Journal of Forestry*, 107, 7 (2009). The Save Leitrim protest was covered by the *Independent*, 30 January 2019 and Agriland, 30 January 2019 (www.agriland.ie/farming-news/trees-are-replacing-people-save-leitrim-protests-outside-dail/). See also *The Economist*, 2 December 2017. Guckian and McCaffrey are profiled in *The Guardian*, 7 July 2019. 'Dark tide of conifers' is from Michael Viney, *The Irish Times*, 10 September 1988. Information on the pearl mussel is from Pearl Mussel Project (www.pearlmusselproject.ie/freshwater-pearl-mussel.html). Hen harriers are discussed in Anthony Caravaggi, 'Forest management and hen harrier circus cyaneus conservation in Ireland', *Irish Birds*, 42 (2020). Peatland forestry is explored in Florence Renou-Wilson and Kenneth A. Byrne, 'Irish peatland forests: lessons from the past and pathways to a sustainable future', in John A. Stanturf (ed.), *Restoration of Boreal and Temperate Forests*, CRC Press (2015) and Irish Government, *National peatlands strategy* (2015). Tiernan estimate is in Dermot Tiernan, 'Environmental and social enhancement of forest plantations on western peatlands', *Irish Forestry*, 64 (2007).

Estimated area of deep peat forest in the Republic is from DAFM, *Ireland's national forest inventory 2017*. The inventory presents figures for stocked areas, which I then adjusted to account for the full forest area. For Northern Ireland, figures for public forests are taken from Forest Research, *Carbon balance of Northern Ireland Forest Service forest on deep peat* (2021). I then assumed the same ratio of public to private forests on peat soil as in the Republic to estimate the total area of private forests on peat soils in Northern Ireland. Of the estimated 346,000 hectares of forestry on deep peat, around 200,000 hectares are older public forests controlled by Coillte or the Northern Ireland Forest Service; the rest are private, mostly younger forests established under the grant schemes. The large area of low-quality private forests is confirmed by the experience of my company, SLM Partners, which reviewed more than 850 properties for sale between 2018 and 2022.

Native woodlands in the Republic are profiled in John Cross, 'Ireland's native woodlands: a summary based on the National Survey of Native Woodlands', *Irish Forestry*, 69 (2012) and Philip M. Perrin and Orla H. Daly, *A provisional inventory of ancient and long-established woodland in Ireland*, Irish Wildlife Manuals No. 46, National Parks and Wildlife Service (2010). Northern Ireland figures are from Woodland Trust, *Back on the map* (ati. woodlandtrust.org.uk/back-on-the-map). Clearing of native woodlands by the Forest Service is described in Margaret Duff Garvey, 'Replanting Ireland: Parliamentary debate and expert literature on Irish state forestry 1922 to 1939', PhD Dissertation, Trinity College Dublin (2020). Similar experience in Northern Ireland is from Woodland Trust (op. cit.). St John's Wood is in Michael Viney, *Ireland: a Smithsonian Natural History*, Smithsonian Books (2010). The Killarney woods are described in Fraser J.G. Mitchell, 'The development of Ireland's tree cover over the millennia', *Irish Forestry* (2000); Viney, *Ireland*; Valerie Hall, 'Pollen analytical investigations of the Irish landscape AD 200–1650', *Peritia*, 14 (2000). The 'lying waste' quote is from Nigel Everett, *The Woods of Ireland: A history, 700–1800*, Four Courts Press (2014) and the Mitchell quote from Fraser J.G. Mitchell, 'Long-term changes and drivers of biodiversity in Atlantic oakwoods', *Forest Ecology and Management*, 307, 1 (2013). Age of broadleaves and planting rates are from DAFM, *Forest Statistics Ireland 2020*. Estimates of the value of forest types per hectare are based on private information, although see also Henry Phillips, 'Economic analysis of broadleaf afforestation', Irish Forest Industry Chain (2005). Prevalence of ash in Irish landscape is from John R. Cross, 'The potential natural vegetation of Ireland', *Biology and Environment: Proceedings of the Royal Irish Academy*, 106B, 2 (2006) and DAFM, *Ireland's national forest inventory 2017*. Ash dieback is analysed in Ian Short and Jerry Hawe, 'Ash dieback in Ireland – a review of European management options and case studies in remedial silviculture', *Irish Forestry*, 75 (2019); A.R. McCracken et al., 'Ash dieback on the island of Ireland', in

Rimvydas Vasaitis and Rasmus Enderle, *Dieback of European Ash*, COST (2017); and the *Independent*, 17 November 2018.

The appeals system is explained in Jo O'Hara, *Implementation of the Mackinnon Report* (February 2021), which is the source of the O'Hara quote. Numbers of appeals and size of backlog are given by thejournal.ie, 25 and 26 October 2020 (www.thejournal.ie/spruced-up-pt1-5241271-Oct2020/), which is the source of the Moran quote. Protests by foresters are described by Agriland on 28 October 2021 (www.agriland.ie/farming-news/failed-and-empty-promises-forestry-group-to-stage-dail-protest).

9. A Sylvan Future?

The story of the founding of Pro Silva and its early leaders is based on this author's interview with Paddy Purser, 19 October 2021. Quote 'his mistrust of authority' and Tottenham biography are from Paddy Purser, 'Obituary: Robert Tottenham', *Irish Forestry*, 64 (2007). Tottenham's younger brother is quoted in *The Irish Times*, 12 May 2007. Jan Alexander is profiled in *The Irish Times*, 23 September 2006 – all her quotes are from this source.

A good summary of continuous cover forestry (CCF), and the quote 'maximise the commercial', are provided by R. Helliwell, *Continuous Cover Management of Woodlands: A brief introduction*, York Publishing Services (2013). Much of this section draws on Paul McMahon et al., *Investing in continuous cover forestry*, SLM Partners White Paper (2016), which summarises research on the economic and environmental case for CCF. A key paper from the UK is O. Davies and G. Kerr, *The costs and revenues of transformation to continuous cover forestry*, UK Forestry Commission (2011). The environmental benefits are also outlined in L. Vítková and A. Ní Dhubháin, 'Transformation to continuous cover forestry: a review', *Irish Forestry*, 70, (2013). Pine weevil is described in W.L. Mason, 'Implementing continuous cover forestry in planted forests: experience with Sitka spruce (*Picea sitchensis*) in the British Isles', *Forests*, 6 (2015). Ash dieback strategy is outlined in Ian Short and Jerry Hawe, 'Ash dieback in Ireland – a review of European management options and case studies in remedial silviculture', *Irish Forestry*, 75 (2019). For carbon figures see Cathal Geoghegan et al., 'The Irish forestry sector', in Cathal O'Donoghue et al. (eds), *Rural Economic Development in Ireland*, Teagasc (2014). Otto's character was described by Paddy Purser in an interview with this author, 19 October 2021. Otto's career and the history of Saxony forestry are from Lars Borrass et al., 'The "German model" of integrative multifunctional forest management – analysing the emergence and political evolution of a forest management concept', *Forest Policy and Economics*, 77 (2017); Hans-Jürgen Otto, *Approaches of close to nature silviculture in Sitka spruce pioneer plantations in Ireland*, Pro Silva Ireland (undated); and 'Prof. Dr. Hans-Jürgen Otto verstorben', *Forst Praxis*, 10 April 2017. Otto's quotes at the prize ceremony are from Wikipedia (de.wikipedia.org/wiki/Hans-J%C3%BCrgen_ Otto). CCF usage today is from W.L. Mason et al., 'Continuous cover forestry in Europe: usage and the knowledge gaps', *Forestry: An International Journal of Forest Research*, 95, 1 (2022). The Danish experience is from a presentation by P. Hilbert to a joint Pro Silva Ireland/IFA conference (2006). For the bark beetle crisis, see 'Climate change to blame as bark beetles ravage central Europe's forests', *Reuters*, 26 April 2019.

The Hayes quote is from Samuel Hayes, *A Practical Treatise on Planting: And the management of woods and coppices* (1794), quoted in T. Clear, 'The role of mixed woods in Irish silviculture', *Irish Forestry*, 1 (1944). Deer damage to woodland is described by the Irish Deer Management Forum (idmf.ie/about-us/). Otto's views on Sitka spruce are in Otto (op. cit.). The over-valuing of 'nativeness' has been explored throughout this book, but this view is also echoed in an interesting recent article by Jim Knight, 'A forest for the future', *Reforesting Scotland*, 61 (2020). North American Sitka spruce forests are profiled in Robert L. Deal et al., 'Lessons from native spruce forests in Alaska: managing Sitka spruce plantations worldwide to benefit biodiversity and ecosystem services', *Forestry: An International Journal of Forest Research*, 87, 2 (2014) and Bernt-Håvard Øyen and Per Holm Nygaard, 'Impact of Sitka spruce on biodiversity in NW Europe with a special focus on Norway – evidence, perceptions and regulations', *Scandinavian Journal of Forest Research*, 35 (2020). British research is in J.W. Humphrey, 'Enhancing biodiversity in UK plantation forests: future perspectives', in L. MacLennan (ed.), *Opportunities for biodiversity enhancement in plantation forests*, Proceedings of COFORD seminar, 24 October 2002, COFORD.

Scottish policy is in Jan Oosthoek, *Conquering the Highlands: A history of the afforestation of the Scottish uplands*, ANU E Press (2013). For backlash against Coillte plans in Dublin Mountains see *The Irish Times*, 25 February 2017. Coillte Nature is described at www.coillte.ie/coillte-nature/coilltenature/ and Coillte's new strategy in 'Coillte launches new forestry strategic vision to optimise its contribution to Ireland's climate targets', Coillte Press Release, 21 April 2022. The case for forestry co-operatives is made at icos.ie/2015/03/23/forestry-co-operatives-are-the-way-to-go/.

INDEX

Act for Planting and Preserving
Timber Trees and Woods
(1698) 93
Addergoole, County Galway
25–6
Adventurers' Act (1642) 81
afforestation 110–12
agriculture and 115, 118, 125
clear-fell rotations 121, 127,
150, 154, 155–6, 164
climate change and 131–2,
148
compartmentalisation 143,
153, 161, 166
decline 146, 156
EEC and 125–6
failed forests 139, 164
forestry licences 114, 136–7,
146, 147
grants 126, 133, 143, 161,
165
Growing for the Future 132
Irish growing conditions 130
legislation 146–7
private landowners 124–5,
126, 128, 130, 131,
144–5, 150, 167
report (1907) 113
state policy 111–12, 114,
115, 116–19, 124–5, 132,
140–1, 143
targets 132–3, 146
see also broadleaf trees; Coillte;
conifers; continuous cover

forestry; Forest Service;
forestry; monocultures
agriculture 18, 19, 23–4, 59, 76,
104
afforestation and 115, 118,
125
Bronze Age 27, 28, 29
forest clearances 20, 22, 23–4,
30, 48–9, 83–4, 85–6, 99,
104
Neolithic period 44, 105
subsidies 124, 133
see also cereal cultivation;
livestock
alder (*Alnus glutinosa*) 5, 10, 38,
46, 48, 54
artefacts 16, 25
climate change, effects of 16,
30
Alexander, Jan 136, 151–2, 153,
157
All Saints Priory, Dublin 62
Allied Irish Bank 126
Anglo-Irish Ascendancy 87, 88
Big House demesnes 89–90,
101, 104
land reforms 106–7, 109,
116, 140
trees 95–6, 101, 102, 125,
151
woodlands, sale of 106–7
Anglo-Normans 37, 54, 55–6
castles 58–9, 61
forest clearances 59, 67

Gaelicisation of 52, 65, 69, 77
livestock 59, 161
manorial system 58–9, 61, 63
towns 60–1
trade 60, 62–3, 65, 66
warfare 57–8
woods, management of
  59–60, 62, 71
Anglo-Saxons 55, 57
Annals of the Four Masters 40
apple/crab apple trees 5, 10, 38,
  41, 46, 141
Aran Islands 104–5
archaeological excavations 18, 27,
  33, 46, 48, 53
ash (*Fraxinus excelsior*) 5, 27–8,
  38, 41, 48, 79, 91
  artefacts 25, 28, 41, 145, 146
  ash dieback 28, 144, 145–6
  construction, use in 54, 74
  Craeb Uisnig (sacred tree)
    33–4
  forestry schemes 95, 144–5
aspen 5, 10, 38
Atlantic Ocean 16
Avondale House and Forest Park
  109–10, 122, 123
  Centenary Trail 110
  forestry school 110–11,
    112–13, 115
axes 19, 43

Balkans 5, 7, 96
Ballykelly Wood, County Derry
  119
Ballynageral, County Waterford
  79
Baltic region 6, 86, 92, 96
Bandon, County Cork 80
Bann River 73, 74–5
Battle of Hastings (1066) 55

bears 10, 15
beech (*Fagus*) 5, 10, 13, 28, 29,
  95–6, 139
  felling 116
  migration 8, 162
  plantations 94, 142
beechnuts 28
Białowieża Forest, Poland 17
biodiversity 5, 6
  EU Habitats Directive 136
  forests and 139, 155, 163,
    168
  hedgerows and 90
bioenergy 132
birch (*Betula*) 2–3, 7, 10, 38, 46,
  87, 142
  Betula pendula 2
  Betula pubescens 2
  charcoal analysis 45
  pollen analysis 48
  wooden trackways 42
birds
  capercaillie 99
  EU Birds Directive 137
  hen harriers 137, 146
  seeds brought by 9
  woodpeckers 128–9
Black Death (bubonic plague) 64
blackberry bushes (*Rubus fruticosus*) 91
blackthorn (*Prunus spinosa*) 5, 10,
  38, 91
Blackwater River 79, 136
Boate, Gerard, *Ireland's Natural
  History* 86
boats/canoes 25–6, 97
Bog of Allen 21
bog bodies 45
bog butter 44
bog myrtle 38
bog wood 21, 27, 92

bogs 137
  blanket bogs 21, 22, 49, 118–19
  carbon storage and 137, 165
  climate change and 23
  formation 20–1
  growth of 20–2, 29, 45, 49
  peatland forests 118–19, 138–9
  pollen, preservation of 6
  raised bogs 21, 49, 119
  restoration projects 164–5
  turf-cutting 92
  warfare and 72, 73
  wooden trackways 25, 42
book (*buch/buche*) 32
Bord na Móna 132
Boyle, Lewis 80–1
Boyle, Richard, Earl of Cork 78–81, 83, 84
Boyne River 34, 61
Brackloon, County Mayo 141
Brehon laws 38, 39, 40, 43, 47–8
Bretha Comaithchesa 38
Brian Boru 26, 34
Britain 29, 130, 163, 164
  ash dieback 145
  forest cover 128, 179
  forestry schools 120
  native trees 5, 7, 8
  Roman Britain 46, 57
  see also England; Northern Ireland; Scotland
broadleaf trees 128, 139, 140, 141, 143, 144, 152
Bronze Age 24–8, 29, 32, 41, 43, 105
Brooks, Anthony 9
Burgh, Thomas 72
Burke, Edmund 88

Burren, the 3, 11–12, 21, 28
Byrne, Gay, *Late Late Show, The* 152

Cabinteely Park, Dublin ix, 129
Cairbre Caitcheann 40
Calendar of State Papers (1601) 72–3
Canary Islands 66, 77, 82
Canterbury Cathedral 62
carbon dioxide xii, 13, 131–2, 137, 155–6, 165
Carbury, County Kildare 34
Carden, Ruth 15
Caribbean 9, 85
Carton, County Kildare 102
Castle Durrow, County Laois 116
Castle Forbes, County Longford 97
Catholic Church 106
Catholics 88, 101–2
  land confiscations 69, 81, 82, 83
  land ownership 82, 83, 89, 102
Cattle Acts (1663, 1667) 85
cattle raids 41, 47–8
Céide Fields, County Mayo 19–20, 22, 23
Celtic languages 31, 32
Celts 31, 32, 34, 35, 36
ceramics 38
cereal cultivation 20, 29, 44, 45, 48, 59, 64
charcoal 79, 80, 83, 86
  analysis 18, 20, 22, 43, 45
chariots 41–2
Christianity 32, 33, 35, 36, 40, 46
City of London Companies 73, 82

Civil Survey (1650s) 100, 140
Clear, Tom 149
climate change 13–14, 23, 24, 30
    1000 AD 63–4
    1000 BC 29
afforestation and 131–2, 148
Climatic Optimum 17
Clonmacnoise, County Offaly
    48, 52
Clonycavan, County Meath 45
Cloragh Farm, County Wicklow
    149, 150
Cloragh Forest, County Wicklow
    151
Coen, Patrick 25
Coillte ix–x, 110, 127, 138–9,
    164, 166–7
    see also Forest Service
Colmcille, St 35, 74
Columbus, Christopher 66
Congested District Board 112
conifers x, xi, 96–8, 101, 120–4,
    135–6, 141
    see also monocultures; Scots
        pine; Sitka spruce
Connemara 21, 112
continuous cover forestry 149,
    150, 151, 153–60, 168
    Coillte and 166
    environmental benefits 155, 156
    Europe and 159
    legislation and 165, 168
    limitations 164
    mixed-age stands, advantages
        of 154, 155
    natural regeneration 153, 154,
        155
    social functions of 156–7
Cooley, County Louth 48
coopers/coopering 41, 82, 86

casks 41, 74, 82, 85, 91, 142
    staves 66, 84, 85, 92, 99, 142
copper mining 12, 26, 28
Cork City 52, 60, 85
Cork County 26, 73, 77, 78,
    80–1, 136
    forests/woods 12, 71, 72, 91,
        98, 113, 141
Corlea Bog, County Longford
    42, 45
Corofin, County Clare 11
Cosby, Philip 102
Craeb Uisnig 33–4
Crann 136, 152
creaght (mixed herd) 65
Cromwell, Oliver xi–xii, 68, 81,
    86, 100
Cross, John 140
Cú Chulainn 33, 47
cypress (*Cupressus*) 11

dairthech (oak-house) 40
Dál gCais sept 34
Dark Ages 32, 45–6, 47
de Moenes, William 62
de-wilding 99, 104
deer 1, 2, 59, 141, 160–1
Deer Park Farm, County Antrim 39
deforestation xi–xii, 22, 29
    see also forest clearances
dendrochronology 18–19, 40, 43,
    51, 62
Denmark 19, 51, 55, 128, 159,
    166, 167
Department of Agriculture 107,
    113, 114
Derry 35
Derryville bog, County Tipperary
    46
Desmond Rebellion 77

Dillon, James 115
Dingle peninsula, County Kerry 44, 46, 49
Domesday survey (1086) 56, 57
Douglas fir (*Pseudotsuga menziesii*) 122
Drogheda 60, 61
Dromana, County Waterford 102–3
Druids 31, 35
Dublin 52, 53, 54, 60, 61, 84
   Baile Átha Cliath (Irish name) 53
   Dubh Linn (Black Pool) 54
   Dyflinn (Old Norse) 53
Dublin Castle 54, 73, 113
Dublin Mountains ix–x, 166
Dublin Society for Improving Husbandry, Manufactures and Other Useful Arts 93–5, 98, 99
Dundaniel, County Cork 78
Dutch elm disease 4, 20, 146

East India Company 78
Easter Rising (1916) 114, 117
ecologists 83, 100, 136, 137, 140, 141, 147, 148
Edward the Bruce 64
Edwards, Robin, and Brooks, Anthony 9
elder 38
Electricity Supply Board (ESB) 122
Elizabeth I, Queen xi–xii, 68, 77, 81–2
Ellis, William, *Timber-tree improved, The* 97–8
elm (*Ulmus glabra*) 4, 7, 10, 30, 38, 48, 74, 95
   Dutch elm disease 4, 20, 146

Emain Macha, County Armagh 33, 40
endangered species 136, 137, 146
England 66, 77, 81, 82, 85
   forest cover 56–7, 71, 100
   forest survey (1796) 92
English rule, decline of 64–5, 66
environmentalists xi, 13, 135–6, 137, 139, 147
Europe
   bogs 23, 164
   forests 17, 29, 46–7, 159
   glaciations 5–6
   native trees 5, 28, 95
   postglacial migration 5, 6, 13
European Economic Community (EEC) 125–6
European Union (EU)
   Birds Directive 137
   Common Agricultural Policy 133
   Directives on climate 131, 132, 137
   Habitats Directive 136, 146
Everett, Nigel 70

famine (1314–22) 63
farmers 106, 107, 125, 130, 133–4, 147
Farrelly, Niall 130
Ferriter's Cove, Dingle, County Kerry 44
Fid Déicsen hi Tuirtre 48
Fid Moithrehi Connachtaib 48
Fid Mór hi Cúailngi 48
fidnemed (tree sanctuary) 35
Finglas, County Dublin 37
Finland 21, 130
Fionn mac Cumhaill 34
firs (*Abies*) 5, 10

First World War (1914–18) 107,
    114, 120
Fitzgerald, Gerald, 14th Earl of
    Desmond 77
Flannery, Revd Thomas 112
Forbes, A.C. 107, 112–13, 114,
    122, 123, 129
forest clearances
    agriculture and 20, 22, 23–4,
        30, 48–9, 83–4, 85–6, 99,
        104
    Anglo-Normans and 59, 67
    Bronze Age 24, 26, 29, 43,
        105
    Iron Age 44–5, 105
    lulls in 24, 45–6, 105
    Neolithic period 23–4
    Ulster planters and 74–6
forest cover in Ireland vi, xi
    1000 AD 49–50
    1170s 63
    1500s 67, 72
    1600s 70–1, 76, 140
    1790s 100, 107
    1800s 100, 140
    1907 107–8, 116
    1950 116
    1980s 119
    2020 128
    Bronze Age 26, 29, 43
    decline 132–3
    ecosystem, emergence of
        128–9
    fluctuations in 67
    Mesolithic period 17, 29
    Neolithic period 24, 29
    old species, return of 128
    UN Report (1950) 116
Forest Service 115, 117, 118,
    119, 120, 124, 127

broadleaf trees 143
failed forests 139, 164, 165
Natura 2000 sites 136–7
projects 137–8
see also Coillte
forestry
    co-operatives 167
    commercial forestry 5, 134–5
    history of 109
    premiums (1766–1806) 94–5,
        98, 99
    science of 119–21, 128, 159
    see also afforestation; Avondale
        House and Forest Park;
        conifers; continuous cover
        forestry; forests; Sitka
        spruce; woods
Forestry Act (1928) 114
Forestry Appeals Committee 147
Forestry Commission 114
Forestry Programmes 126, 133, 161
forests 2, 48, 69–70, 76, 89, 142
    biodiversity and 139, 168
    carbon dioxide and xii,
        131–2, 155–6
    interglacial periods 10–11
    planters, exploitation by
        73–4, 75, 76, 79–80, 82
    primeval 17, 48, 49
    regeneration of 24, 45
    secondary forests 29, 46
fossils 6, 11, 12, 21, 27, 92, 165
France 5, 7, 17, 66, 82, 85, 161
fuchsia (*Fuchsia magellanica*) 91
fulachtí fiadha 25
fungi 1, 17, 163

Gaelic Irish 52, 59, 74, 79, 81
Galway-Mayo Institute of
    Technology (GMIT) 148

Geoffrey of Monmouth 56
Gerald of Wales 36–7, 56, 57, 86
    Topographia Hibernica 56
Germany 100, 119–20, 150,
    157–9, 161, 166
Gillespie, Raymond 84
Glanageenty Wood, County Kerry
    77
glassworks 79, 82
Glendalough, County Wicklow
    35–6, 49
Glengariff Woods, County Cork
    141
Glenglas Forest (Cork–Limerick
    border) 71
Glenkonkyne Forest, County
    Derry 73, 74, 75, 76, 77, 82
gorse 38, 91
Great Famine 105
Great Fire of London (1666) 82
Green Party (German) 158
Green Party (Irish) x
greenhouse gases 131–2, 168
Greenland 29, 64
Guckian, Edwina 135
guelder-rose (*Viburnum opulus*)
    91
Gulf Stream 9

Hall, Valerie 17, 22, 70
Harold II, King 55
hawthorn (*Crataegus monogyna*)
    5, 10, 38, 91, 108
Hayes, Samuel, *Practical Treatise
    on Planting, A; and The
    Management of Woods and
    Coppices* 109, 160
hazel (*Corylus avellana*) 3, 5, 7,
    10, 20, 91
    artefacts 25

Celtic mythology 34, 39
    construction, use in 25, 39, 54
    fines for cutting 38
    pollen analysis 39, 48
    woodland 39, 46, 57, 64,
        76–7, 141
hazel mead 39
hazelnuts 3, 16, 33, 34, 38, 39,
    42, 76
heather 19, 20, 21, 38, 137
Hecataeus of Miletus 31
hedgerows 90–1, 95, 100, 145
Hen Harrier Special Protection
    Areas (SPAs) 137
Henry II, King 55, 56, 61
Henry III, King 61
Henry VIII, King 67, 69
Hiberno-Norse 52, 55
High Kings of Ireland 39
highways/roads 41–2, 53, 58
    see also wooden trackways
Hogan, Patrick 111
holly (*Ilex aquifoloum*) 5, 10, 38,
    41, 46
Hollywood, County Wicklow 35
Holocene period 9, 10, 14
hornbeam 10, 13
horse chestnut (*Aesculus hippocas-
    tanum*) 95, 96
houses 25, 61, 65
    wattle walls 39, 54, 93
Howitz, Daniel, *Reafforesting of Waste
    Lands in Ireland, The* 112
Hvid, Carsten 51

Iberian Peninsula 12
Ice Age xi, 1, 7–8
    interglacial periods 10–11
    postglacial migration 5–6, 13,
        28, 105, 162

immigrants 58, 63, 69, 75, 84
Inchagreenoge, County Limerick
   27
Indian Forest Service 120
insects 1, 7, 17, 90, 123, 163
International Dark Sky Park 24
Ireland
   formation of 11
   population (1000–500 BC)
      29
   population (1300) 63
   population (1348–9) 64
   population (1500) 65
   population (1550–1641) 84
   population (1730–1845) 103
   population decline 105
Iria the Prophet 49
Irish Act (1297) 58
Irish Farmers Association (IFA)
   135
Irish Free State 114
Irish language 30, 32
Irish Life 126
Irish Peatland Conservation
   Council 11
Irish Republican Army (IRA)
   114, 117
Irish Sea 7, 8, 53
Iron Age 30, 31–2, 46
   chariots 41–2
   forest clearances 44–5, 105
   La Tène art 31, 32
   Late Iron Age Lull 45
   wood, importance of 37–8
ironworks 80, 82, 83, 84, 86, 99,
   142
Italy 5, 7, 17, 27, 31

James I, King 75
James II, King 82

Joyce, James, *Ulysses* 110–11
Julius Caesar 57–8
juniper (*Juniperus communis*) 1,
   10, 15, 26, 38, 139

Kelly, Eamonn 26
Kerry slug 12
Kevin, St 35–6, 49
Kilbegly, County Roscommon 48
Kilgarvan, County Kerry 151
Killarney National Park 27, 49,
   141–2
Killetra Forest, County Derry 73,
   74, 75, 76, 82
Knights Hospitallers 141
Knockboy, Connemara 112
Knockrath, County Wicklow 151
Knowth, County Meath 18

La Tène art 31, 32
Lambay Island 46
Land Acts 106, 110, 114, 125
land bridges 7, 8–9
Land Commission 114–15, 125,
   140
land confiscations 69, 81, 82–3
   see also plantations/planters
land reform 106–7, 108, 109,
   116, 125
landnam (forest clearances) 20
larch (*Larix decidua*) 97
Laserian, St 36
Last Glacial Maximum 1
Leitrim, Save Leitrim Campaign
   135–6, 139, 147
Lewis, Samuel, *Topographical
   Dictionary of Ireland, A* 101
Liberty Trees 103
lichens 1, 27, 163
Liffey River 53, 54

lime trees 8, 10, 13
Limerick 52, 60, 66
limestone areas 17, 28, 39, 91, 106
  see also Burren, the
Lismore Castle, County Waterford
  79
Little Ice Age (1350–1850) 64
livestock 59, 65, 105–6
  cattle 44, 59, 65, 84–5, 106
  cattle plague (1315–16) 63
  cows 44, 59, 131
  pigs 40, 59
  sheep 59, 65, 85, 161
Lloyd George, David 107
lodgepole pine (*Pinus contorta*)
  122–3, 138
Londonderry 74–5
Lough Neagh 71, 73
Lough Ree, County Roscommon
  141
LÖWE programme 159
'Lusitanian' species 12

MacBride, Seán 116–17, 129
McCaffrey, Jim 135
McCracken, Eileen, *Irish Woods
  Since Tudor Times, The* 70–1,
  76, 98–9, 106–7
McEvoy, T. 125
Mackay, John
  No Forests, No Nationhood 69
  Rape of Ireland, The 69
Mac Murchada, Aífe 55
Mac Murchada, Diarmuid, King
  of Leinster 55
MacMurrough Kavanagh, Art,
  King of Leinster 58
Madden, Revd Samuel, *Reflections
  and resolutions proper for the
  gentlemen of Ireland* 93–4

Madeira 66, 77
Magh Adair, Quin, County Clare
  34
Malta 128
Marshall, William 92
Medb, Queen 47
Mesolithic period 16, 17, 18, 29,
  44
Middle Ages 33, 37, 39, 48
Mitchell, Fraser J.G. 23, 142
monasteries 35, 36, 39, 48, 52,
  53
monkey-puzzle (*Araucaria*) 11,
  109
monocultures 120, 134, 153,
  155, 156, 158, 160
Moran, James 148
mosses 1, 20, 21, 27, 39, 61
Mount Callan, County Clare
  149–50
Mount Leinster, County Carlow
  99
Mount Sandel, County Derry 16
Munster plantation 77–8, 80, 82

Napoleonic Wars 100, 142
National Museum of Ireland
  (NMI) 15, 25–6
National Parks and Wildlife
  Service 140
National Survey of Native
  Woodlands 140, 143
native tree species xi, 2–5, 6, 8,
  26, 139–40
  conifers 26, 96–7, 139–40
  definition 13, 162
  human introductions 10, 11,
  12, 94–6, 139
Native Woodland Scheme 11
Native Woodland Trust 136

Natura 2000 sites 136–7
Navan Fort (*Emain Macha*),
  County Armagh 33, 40
Neolithic period 18, 19, 23–4,
  27–9
  agriculture 44, 105
  field system 19–20, 22, 23
Netherlands 128
Newfoundland spruce 95
Newgrange, County Meath 18
Nicholls, Kenneth 71
Nine Years' War 72–3
Normans 55, 57
  see also Anglo-Normans
Norsemen 52, 55
North America 5, 6, 28, 85, 92,
  122–3, 163
Northern Ireland 119, 125, 127,
  140, 141
Norway spruce (*Picea abies*) 97,
  124, 157–8

oak (*Quercus*) 10, 38, 39, 42, 45,
  64
  acorns 4, 40
  artefacts 25–6, 66
  bark 39–40, 79, 85, 92
  Brehon law and 39, 40
  construction, use in 25, 40,
    54, 74, 86
  DNA analysis of 7
  doire (oak grove) 35, 74
  Druids and 31, 35
  exports 62, 66–7
  felling 74, 75–6, 99, 142
  forestry schemes 95
  growth rate 144
  migration of 7
  pedunculate oak (*Quercus
    robur*) 4

primeval forest 17
sessile oak (*Quercus petraea*) 4
tree-ring analysis 19
value 91, 144
woodland 35, 141, 142
O'Connell, Daniel 106
O'Donnell, Red Hugh 72
O'Dowd, Marion 15
O'Hara, Jo 147
Old English 69, 79, 81, 82
Old Leighlin, County Carlow 36
O'Neill, Hugh, Earl of Tyrone
  72, 73, 77
Ordnance Survey (1830s) 98,
  100, 140
Otto, Hans-Jürgen 157–8, 162
O'Tuama, Padraig 151

paganism 33, 35, 36
Pale, the 65, 71
paleo-botanical research 20
paleo-ecologists 6, 8, 9
paleo-geologists 6, 8, 9
Palaeolithic period 15
Palladius, St 46
Parnell, Charles Stewart 106, 109
Patrick, St 35, 46, 48
peatland forests 118–19, 138–9
penal laws 89
Perrin, Philip M. 140
Pilcher, Jonathan, and Hall,
  Valerie, *Flora Hibernica* 17,
  22
pine 7, 15, 20, 23, 24, 38, 122
  see also Scots pine
plantations/planters 69–70, 88–9
  Cromwellian 81, 82
  forests, exploitation of 73–4,
    75, 76, 79–80, 82, 89
  Munster 77–8, 82

Survey of the plantation of Munster 82
Ulster 73–5
Pleistocene period 10
Pliny 35
Poddle River 53, 54
pollen analysis 6, 12–13, 18, 20, 44
  alder 16
  Aran Islands 105
  ash 28
  cereal cultivation 20, 48, 63
  elm 20
  hazel 39
  hornbeam 13
  Iron Age 45, 46
  Killarney woodlands 142
  lime 13
  oak 75–6
  ribwort 29
  Scots pine 12
  walnut 28
  woodland decline 45, 49, 63
Pomponius Mela 44
Portglenone Forest 75
Portugal 66
potato blight (*Phytophthora infestans*) 105
potatoes 103–4, 106
Powerscourt, County Wicklow 95, 122, 141
pre-Christian Ireland 33, 35
Pro Silva Europe 149, 150
Pro Silva Ireland 149, 150–1, 152–3, 157, 166
Programme for Economic Development 118
Protestant landlords 37, 90, 102–3
  see also Anglo-Irish Ascendancy

Public Records Office 70
Purser, Paddy 151

Queen's University Belfast (QUB) 19, 70

rabbits 59
Rackham, Oliver 83, 100
Raleigh, Sir Walter 77–8, 79, 83
Rathlin Island 119
ráths (ringforts) 47
redwood (*Sequoia*) 109, 123, 163
Reenadinna Wood, Killarney National Park 27
reforestation xii, xiii, 108, 111
  see also afforestation
Reinhard, Otto 120
rhododendron 142
ribwort (*Plantago lanceolata*) 29
Richard II, King 58
ring-barking 19
ringforts (*ráths*) 47
Roche, Jenni 11–12
Roche, Morgan 151
Rockforest Wood, County Clare 11–12, 21, 96
Roman Britain 46, 57, 58
Roman Empire 31, 46, 47
Roman literature 31, 35
Roskilde Fjord, Denmark 51
rowan 5, 10, 38
Royal Dublin Society (RDS) 93
Royal Navy 68, 92

sacred trees xi, 33–4, 108
  bile (sacred tree) 35
  Christianity and 35, 36, 37
  Craeb Uisnig 33–4
  Druids and 31, 35, 36
  Eó Rossa (yew) 36

fidnemed (tree sanctuary) 35
hazel trees 34
inauguration sites 34, 48
St Kevin's Yew 36
slat na ríghe 34
symbolism 34
taboos 37, 50, 108
Toothache Tree 37
veneration of 36, 37, 108
yew 36, 40
St John's Wood, County
  Roscommon 141
St Patrick's Cathedral, Dublin 61,
  92
Salisbury Cathedral 62
Scotland 65–6, 82, 100, 119, 165
Scots pine (*Pinus sylvestris*) 3–4,
  10, 26–7, 139–40
  bogs and 21, 30
  felling 142
  origin of 11, 96–7, 162
  relict stands of 11–12, 96
Sea Stallion from Glendalough
  (longship) 51
sedges 1, 20
Shakespeare, William, *Henry IV*
  74
Sheehy Skeffington, Micheline,
  and Scott, Nick 12
Shillelagh, County Wicklow 66,
  99
shipbuilding 51, 54–5, 61, 78,
  81, 84
silver fir (*Abies alba*) 97
silvo-pastoralism 77
Sitka spruce (*Picea sitchensis*) ix–x,
  122, 128, 130
  biodiversity and 163
  German plantations 157–8
  growing conditions 135, 138

growth rate 143–4
management 160, 162, 163–4
natural regeneration 160
plantations, opposition to 135
private landowners and 150
value of 144
Sitriuc Silkenbeard 52–3
Skuldelev 2 (Viking longship) 51,
  55
slat na ríghe (rod of kingship) 34
Slieve Gallion, County Tyrone 48
Slíghe Cualann (highway) 53
Sligo County 12
Sluggan Moss, County Antrim 49
smallholders 104, 106
Smith, Charles 98
soils 1, 21, 26–7, 46, 137, 138
Somerset Levels, England 27
Spain 5, 7, 28, 40, 73
  trade with 66, 67, 82, 85
Spanish Arch, Galway 66
spindle (*Euonymus europaeus*) 38,
  91
spruces (*Picea*) 5, 10, 95, 97
  see also Sitka spruce
squirrels 99–100, 163
Stolze, Susann, and Monecke,
  Thomas 12–13
Stone Age 19
  see also Mesolithic period;
    Neolithic period;
    Palaeolithic period
Stradbally, County Laois 102
strawberry tree (*Arbutus unedo*) 5,
  7, 10, 12, 28, 38
Strongbow 55, 62, 63
Sweden 130, 167
Swift, Jonathan
  Drapier's Letters 92–3
  Gulliver's Travels 92

Switzerland 150, 151, 159
sycamore (*Acer pseudoplatanus*) 5, 13, 28–9, 30, 95, 139

Tacitus 31
Táin Bó Cúailnge 41, 47
tanners/tanning 39, 40, 79, 82, 84, 85, 86, 92, 142
Tara, County Meath 33, 39, 53
Teagasc 130
Terenure (*Tír an Iúir*) 27
Tiernan, Dermot 138, 139
timber
    exports 52, 53, 62, 66–7, 77–8, 82, 85, 87, 99, 106
    imports 86, 87, 92, 94, 96
    scarcity 91, 92, 101, 107
timber-processing industry 118, 123, 127, 130–1, 147
    bioenergy, impact of 132
    exports 131
    medium-density fibreboard (MDF) 123, 131
    oriented strand board (OSB) 131
    pulpwood 123, 127, 131, 132
    sawmills 109, 118, 123, 127, 130–1, 147
Tomies Wood, County Kerry 142
Tomnafinnoge Wood, County Wicklow 99
Toothache Tree, Beragh Hill, County Tyrone 37
Tottenham, Robert 149–51, 153
Tree Council of Ireland 136
tree seeds, vectors and 9, 162
tree-ring analysis 18–19, 40, 86
trees
    Airig Fedo 38, 39, 40–1
    Aithig Fedo 38

appreciation of 168–9
Bretha Comaithchesa and 38
categories 38
exotic species 10, 11, 95, 101, 109, 121–2
Fodla Fedo 38
Losa Fedo 38
pioneer species 2, 97
planting legislation 93, 100
planting premiums 94–5
see also native tree species; sacred trees
Trefuilngid Tre-eocha 33
Trinity College Dublin (TCD) 9, 11–12, 23, 62, 88, 142
Tudor conquest of Ireland 67, 69, 72, 77
Tyrolean Iceman ('Ötzi') 27

Uisneach, County Westmeath 33–4
Ulster 34, 47, 72, 73–6
Ulster Cycle 33
UN Food and Agriculture Organization 116, 117
United Irishmen 103
United Kingdom (UK) 127
United States of America (USA) 5, 28
urban development 52, 53, 60–1, 80, 84

Viking Ship Museum, Roskilde 51
Vikings 26, 51–4, 66
Vita Prima Sanctae Brigitae 43

walnut (*Juglans*) 28, 29, 94
warfare 29, 57–8, 72–3, 77
Waterford 52, 54, 55, 60, 66

Watson-Wentworth family 99
Westminster Hall, London 99
Wexford 52, 55, 60, 66
Whitaker, T.K. 118
whitebeam (*Sorbus*) 5, 10, 38, 141
Whiteboys 102–3
wild cherry 38
wild rose 38
Wilhelm Leopold Pfeil Prize 158
William the Conqueror 55
Williamite Wars 82, 86, 88, 142
willow (*Salix*) 10, 38, 46, 51, 91
Wolfwalkers (film) 68
wolves 68, 74, 80, 99, 161
wooden trackways 25, 42, 45
woodkernes 74, 80
Woodland League 136
woodmen/woodmanship 43, 60, 107, 108
woods 10
    ancient 140, 141

coppicing 39, 42, 60, 67, 83, 86
damp woodlands 16
dark wood 17
deer, overgrazing by 142, 160–1
felling 81–2, 93, 116, 141
management 42–3, 59–60, 62, 83, 141
regeneration 64, 67, 105, 142, 153, 154
rod-cutting project 62
sale of 98–9
see also forests

yew (*Taxus Baccata*) 5, 10, 26–7, 30, 38, 46, 139
artefacts 25, 27, 40–1, 54
decline of 27, 48
Eó Rossa (sacred tree) 36
St Kevin's Yew 36
Young, Arthur 100, 102